SCHAUM'S OUTLINE OF

Theory and Problems of
Astronomy

STACY E. PALEN

Department of Astronomy
University of Washington

Schaum's Outline Series

McGRAW-HILL

New York Chicago San Francisco Lisbon London Madrid Mexico City
Milan New Delhi San Juan Seoul Singapore Sydney Toronto

STACY E. PALEN, currently a lecturer and research scientist at the University of Washington, received her Ph.D. from the University of Iowa in stellar astrophysics. During her graduate career, she authored a unique set of laboratory exercises for introductory astronomy classes, utilizing remote-controlled telescopes over the internet. Subsequently, in addition to teaching college-level astronomy courses at the University of Washington, she compiles and maintains a diverse clearinghouse of astronomy activities on the web. At the behest of several of her students, she also founded the Undergraduate Astronomy Institute, an innovative learning environment for undergraduate students to explore independent astrophysics research. Dr. Palen is nationally recognized both for her research and for her continued contributions to astronomy education.

Schaum's Outline of Theory and Problems of
ASTRONOMY

Copyright © 2002 by The McGraw-Hill Companies, Inc. All rights reserved. Printed in the United States of America. Except as permitted under the Copyright Act of 1976, no part of this publication may be reproduced or distributed in any form or by any means, or stored in a data base or retrieval system, without the prior written permission of the publisher.

12 13 14 15 DIG/DIG 16 15 14

ISBN 0-07-136436-6

Sponsoring Editor: Barbara Gilson
Production Supervisor: Elizabeth J. Shannon
Editing Liaison: Maureen B. Walker
Project Supervision: Keyword Publishing Services Ltd.

Library of Congress Cataloging-in-Publication Data applied for.

McGraw-Hill
A Division of The McGraw-Hill Companies

PREFACE

The purpose of this Outline is to serve as a supplement to a basic astronomy text. Much of the material here is abbreviated, and students should use this book as a guide to the key concepts in modern astronomy, but not as an all-inclusive resource.

Topics covered range from planetary astronomy to cosmology, in the modern context. The first chapter covers most of the phsyics required to obtain a basic understanding of astronomical phenomena. The student will most likely come back to this chapter again and again as they progress through the book. The order of the topics has been set by the most common order of these topics in textbooks (near objects to far objects), but many of the chapters are quite independent, with few references to previous chapters, and may be studied out of order.

The text includes many worked mathematical problems to support the efforts of students who struggle particularly in this area. These detailed problems will help even mathematically adept students to see how to solve astronomical problems involving several steps.

I wish to thank the many people who were instrumental to this work, including the unknown reviewer who gave me so many useful comments, and especially the editors, Glenn Mott of McGraw-Hill and Alan Hunt of Keyword Publishing Services Ltd., who guided a novice author with tremendous patience. I also wish to thank John Armstrong for proofreading the very first (and therefore very rough!) draft and all my colleagues at the University of Washington who served as sources of knowledge and inspiration.

STACY PALEN

CONTENTS

CHAPTER 1	**Physics Facts**	**1**
	About Masses	1
	About Gases	13
	About Light	15
	About Distance	25
CHAPTER 2	**The Sky and Telescopes**	**31**
	Coordinate Systems and Timescales	31
	Instrumentation	42
CHAPTER 3	**Terrestrial Planets**	**49**
	Formation of Terrestrial Planets	50
	Evolution of the Terrestrial Planets	53
	Mercury	56
	Venus	59
	Earth	62
	Moon	67
	Mars	71
	Moons of Mars	72
CHAPTER 4	**Jovian Planets and Their Satellites**	**77**
	Jupiter	78
	Saturn	79
	Uranus	79
	Neptune	80
	Moons	81
	Rings	87
CHAPTER 5	**Debris**	**91**
	Comets	91
	Meteorites	95
	Asteroids	99
	Pluto and Charon	103
CHAPTER 6	**The Interstellar Medium and Star Formation**	**107**
	The Interstellar Medium	107
	Star Formation	117

CHAPTER 7	**Main-Sequence Stars and the Sun**	**125**
	Equilibrium of Stars	125
	Observable Properties of Stars	126
	The Sun	137

CHAPTER 8	**Stellar Evolution**	**153**
	Why Do Stars Evolve?	153
	How Do Stars Evolve?	156
	Stars < $8 M_{Sun}$	158
	Stars > $8 M_{Sun}$: Supernovae	162
	Where Do We Come From?	163

CHAPTER 9	**Stellar Remnants (White Dwarfs, Neutron Stars, and Black Holes)**	**173**
	Degenerate Gas Pressure	173
	White Dwarfs	174
	Neutron Stars	176
	Black Holes	179

CHAPTER 10	**Galaxies and Clusters**	**183**
	The Milky Way	183
	Normal Galaxies	189
	Active Galaxies and Quasars	196

CHAPTER 11	**Cosmology**	**203**
	Hubble's Law	203
	Hubble's Law and the Expansion of the Universe	204
	Hubble's Law and the Age of the Universe	204
	Hubble's Law and the Size of the Universe	205
	The Big Bang	205
	Life in the Universe	214

APPENDIX 1	**Physical and Astronomical Constants**	**219**
APPENDIX 2	**Units and Unit Conversions**	**221**
APPENDIX 3	**Algebra Rules**	**223**
APPENDIX 4	**History of Astronomy Timeline**	**225**
INDEX		**231**

CHAPTER 1

Physics Facts

See Appendix 1 for a list of physical and astronomical constants, Appendix 2 for a list of units and unit conversions, Appendix 3 for a brief algebra review, and the tables in Chapters 3 and 4 for planetary data, such as masses, radii, and sizes of orbits.

About Masses

MASS

Mass is an intrinsic property of an object which indicates how many protons, neutrons, and electrons it has. The *weight* of an object is a force and depends on what gravitational influences are acting (whether the object is on the Earth or on the Moon, for example), but the *mass* stays the same. Mass is usually denoted by either m or M, and is measured in kilograms (kg).

VOLUME

The volume of a body is the amount of space it fills, and it is measured in meters cubed (m^3). The surface of a sphere is $S = 4\pi r^2$ and its volume $V = 4/3\pi r^3$, where r is the radius and π is 3.1416.

DENSITY

The density of an object, the ratio of the mass divided by the volume, is often depicted by ρ (Greek letter 'rho'): it is usually measured in kg/m^3:

$$\rho = \frac{m}{V}$$

GRAVITY

Gravity is the primary force acting upon astronomical objects. Gravity is always an attractive force, acting to pull bodies together. The force of gravity between two homogeneous spherical objects depends upon their masses and the distance between them. The further apart two objects are, the smaller the force of gravity between them. The gravity equation is called Newton's law of gravitation:

$$F = \frac{G \cdot M \cdot m}{d^2}$$

where M and m are the masses of the two objects, d is the distance between their centers, and G is the gravitational constant: 6.67×10^{-11} m^3/kg/s^2. The unit of force is the newton (N), which is equal to $1 \,\text{kg} \cdot \text{m/s}^2$. If the sizes of the two objects are much smaller than their distance d, then the above equation is valid for arbitrary shapes and arbitrary mass distributions.

THE ELLIPSE

The planets orbit the Sun in nearly circular elliptical orbits. An ellipse is described by its major axis (length = $2a$) and its minor axis (length = $2b$), as shown in Fig. 1-1. For each point A on the ellipse, the sum of the distances to the foci AF and AF$'$ is constant. More specifically:

$$\text{AF} + \text{AF}' = 2a$$

The eccentricity of the ellipse is given by $e = \text{FF}'/2a$. In terms of the semi-major axis, a, and the semi-minor axis, b, we have

$$e = \sqrt{1 - \left(\frac{b}{a}\right)^2}$$

The Sun occupies one of the foci in the ellipse described by a planet. Assume that the Sun occupies the focus F. When the planet is on the major axis and at the point nearest F, then the planet is at perihelion. On the far point on the major axis, the planet is at aphelion. By definition,

$$d_p + d_a = 2a$$

where d_p is the planet–Sun distance at perihelion and d_a is the distance at aphelion.

CHAPTER 1 Physics Facts

KEPLER'S LAWS

Kepler's First Law. Planets orbit in ellipses, with the Sun at one focus.

Kepler's Second Law. The product of the distance from the focus and the transverse velocity is a constant. The transverse velocity is the velocity perpendicular to a line drawn from the object to the focus (see Fig. 1-1). As a hint for working problems, consider that when a planet is at aphelion or perihelion (farthest and closest to the Sun), *all* of the velocity is transverse. An alternative statement is that the line from the planet to the Sun sweeps out equal areas in equal periods of time.

Kepler's Third Law. The ratio of the square of the period, P (the amount of time to compete one full orbit), and the cube of the semi-major axis, a, of the orbit is the same for all planets in our solar system. When P is measured in years, and a in astronomical units, AU (1 AU is the average distance from the Earth to the Sun), then Kepler's Third Law is expressed as

$$P^2 = a^3$$

Using Newtonian mechanics, Kepler's Third Law can be expressed as

$$P^2 = \frac{4\pi^2 a^3}{G(m+M)}$$

where m and M are the masses of the two bodies. This Newtonian version is very useful for determining the masses of objects outside of our solar system.

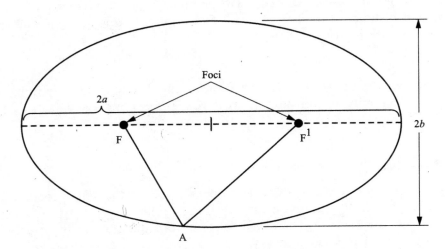

Fig. 1-1. An ellipse. The object being orbited, for example, the Sun, is always located at one focus.

CIRCULAR VELOCITY

If an object moves in a *circular* orbit around a much more massive object, it has a constant speed, given by

$$v_c = \sqrt{\frac{GM}{d}}$$

where M is the mass of the body in the center and d is the distance between the objects. When the orbit is elliptical, rather than circular, this equation is still useful—it gives the *average* velocity of the orbiting body. Provided that d is given in meters, the units of circular velocity are m/s (with $G = 6.67 \times 10^{-11}$ m^3/kg/s^2).

ESCAPE VELOCITY

An object of mass m will remain in orbit if its speed at distance d does not exceed the value

$$v_e = \sqrt{\frac{2GM}{d}}$$

the so-called escape velocity. Again, M is the mass of the larger object. The escape velocity is independent of the mass of the smaller object, m.

ANGULAR MOMENTUM

All orbiting objects have a property called **angular momentum**. Angular momentum is a conserved quantity. Changing the angular momentum of a system requires external action (a net torque). The angular momentum depends on the mass, m, the distance from the object it is orbiting, r, and the transverse velocity, v, of the orbiting object,

$$L = m \cdot v \cdot r$$

The mass of planets is constant, so conservation of angular momentum requires that the product $v \cdot r$ remains constant. This is Kepler's Second Law.

KINETIC ENERGY

Moving objects have more energy than stationary ones (at the same potential energy, for example, at the same height off the ground). The energy of the motion is called the kinetic energy, and depends on both the mass of the object, m, and its velocity, v. The kinetic energy is given by

$$\mathrm{KE} = \tfrac{1}{2}mv^2$$

CHAPTER 1 Physics Facts

Kinetic energy (and energy in general) is measured in joules (J): 1 joule $= 1\,\text{kg} \cdot \text{m}^2/\text{s}^2$. **Power** is energy per unit time, commonly measured in watts (W), or J/s.

GRAVITATIONAL POTENTIAL ENERGY

Gravitational potential energy is the energy due to the gravitational interaction. For two masses, m and M, held at a distance d apart,

$$E_g = -\frac{GmM}{d}$$

Solved Problems

1.1. Assume that your mass is 65 kg. What is the force of gravity exerted on you by the Earth?

Use Newton's law of gravitation,

$$F = \frac{GmM}{d^2}$$

The mass of the Earth is given in Appendix 2, 5.97×10^{24} kg, and the radius of the Earth is 6,378 km (i.e., 6,378,000 m or 6.378×10^6 m). Plugging all of this into the equation gives

$$F = \frac{6.67 \times 10^{-11}\,\text{m}^3/\text{kg}/\text{s}^2 \cdot 65\,\text{kg} \cdot 5.97 \times 10^{24}\,\text{kg}}{(6.378 \times 10^6\,\text{m})^2}$$

$$F = 636\,\text{kg} \cdot \text{m}/\text{s}^2$$

$$F = 636\,\text{N}$$

The gravitational force on you due to the Earth is 636 newtons. This is also the gravitational force that you exert on the Earth. (Try the calculation the other way if you don't believe this is true.)

1.2. What is the maximum value of the force of gravity exerted on you by Jupiter?

The maximum value of this force will occur when the planets are closest together. This will happen when they are on the same side of the Sun, in a line, so that the distance between them becomes

$$d = (d_{\text{Sun to Jupiter}}) - (d_{\text{Sun to Earth}})$$
$$d = 5.2\,\text{AU} - 1\,\text{AU}$$
$$d = 4.2\,\text{AU}$$

Convert AU to meters by multiplying by 1.5×10^{11} m/AU, so that the distance from the Earth to Jupiter is 6.3×10^{11} m. Suppose that your mass is 65 kg, as in Problem 1.1. Therefore, the force of gravity between you and Jupiter is

$$F = \frac{GmM}{d^2}$$
$$F = \frac{6.67 \times 10^{-11}\,\text{m}^3/\text{kg/s}^2 \cdot 65\,\text{kg} \cdot 2 \times 10^{27}\,\text{kg}}{(6.3 \times 10^{11}\,\text{m})^2}$$
$$F = 2.2 \times 10^{-5}\,\text{kg} \cdot \text{m}^2/\text{s}^2$$
$$F = 2.2 \times 10^{-5}\,\text{N}$$

The gravitational force between you and Jupiter is 2.2×10^{-5} newtons.

1.3. What is the gravitational force between you and a person sitting 1/3 m away? Assume each of you has a mass of 65 kg. (For simplicity, assume all objects are spherical.)

$$F = \frac{GmM}{d^2}$$
$$F = \frac{6.67 \times 10^{-11}\,\text{m}^3/\text{kg/s}^2 \cdot 65\,\text{kg} \cdot 65\,\text{kg}}{(0.3\,\text{m})^2}$$
$$F = 3.1 \times 10^{-6}\,\text{N}$$

This is only a factor of about 7 less than the gravitational force due to Jupiter calculated in the previous problem. Despite Jupiter's large size, it would take only 7 people in your vicinity to have a larger gravitational effect on you.

1.4. If someone weighs (has a gravitational force acting on them) 150 pounds on Earth, how much do they weigh on Mars?

The most obvious way to work out this problem is to calculate the person's mass from their weight on Earth, then calculate their weight on Mars. However, many of the terms in the gravity equation are the same in both cases (G and the mass of the person, for example). If you set up the ratio immediately, by dividing the two equations, the calculation is simplified. It is important in this method to put subscripts on all the variables, so that you can keep track of which mass is the mass of Mars, and which radius is the radius of the Earth.

Dividing the equations for the weight on Mars and the weight on Earth gives

$$\frac{F_{\text{Mars}}}{F_{\text{Earth}}} = \frac{\frac{GmM_{\text{Mars}}}{r^2_{\text{Mars}}}}{\frac{GmM_{\text{Earth}}}{r^2_{\text{Earth}}}}$$

The factors of G and m cancel out, so that the equation simplifies to

CHAPTER 1 Physics Facts

$$\frac{F_{Mars}}{F_{Earth}} = \frac{\frac{M_{Mars}}{r_{Mars}^2}}{\frac{M_{Earth}}{r_{Earth}^2}}$$

$$\frac{F_{Mars}}{F_{Earth}} = \frac{M_{Mars} \cdot r_{Earth}^2}{M_{Earth} \cdot r_{Mars}^2}$$

$$\frac{F_{Mars}}{F_{Earth}} = \frac{6.39 \times 10^{23} \text{ kg} \cdot (6{,}378 \text{ km})^2}{5.07 \times 10^{24} \text{ kg} \cdot (3{,}394 \text{ km})^2}$$

$$\frac{F_{Mars}}{F_{Earth}} = \frac{2.6 \times 10^{31}}{5.8 \times 10^{31}} = 0.45$$

The weight of a person on Mars is about 0.45 times their weight on the Earth. For a person weighing 150 pounds on Earth, their weight on Mars would decrease to $0.45 \times F_{Earth} = 0.45 \times 150 = 67$ pounds. Working the problem in this way enables you to skip steps. You do not need to find the mass of the person on the Earth first, and you do not need to plug in all the constants, since they cancel out.

1.5. What is the circular velocity of the space shuttle in lower Earth orbit (300 km above the surface)?

In the circular velocity equation, M is the mass of the object being orbited—in this case, the Earth—and d is the distance between the *centers* of the objects. Since G is in meters, and our distance is in kilometers, convert the distance between the space shuttle and the center of the Earth to meters:

$$d = R_{Earth} + h_{Orbit}$$
$$d = 6{,}378 + 300 \text{ km}$$
$$d = 6{,}678 \text{ km} \cdot \frac{1{,}000 \text{ m}}{\text{km}}$$
$$d = 6{,}678{,}000 \text{ m}$$
$$d = 6.678 \times 10^6 \text{ m}$$

Now use the circular velocity equation:

$$v_c = \sqrt{\frac{GM_{Earth}}{d}}$$

$$v_c = \sqrt{\frac{6.67 \times 10^{-11} \text{ m}^3/\text{kg/s}^2 \cdot 5.97 \times 10^{24} \text{ kg}}{(6.678 \times 10^6 \text{ m})}}$$

$$v_c = \sqrt{5.96 \times 10^7 \frac{\text{m}^3 \cdot \text{kg}}{\text{m} \cdot \text{kg} \cdot \text{s}^2}}$$

$$v_c = \sqrt{5.96 \times 10^7 \frac{\text{m}^2}{\text{s}^2}}$$

$$v_c = 7.72 \times 10^3 \text{ m/s}$$

$$v_c = 7.72 \text{ km/s}$$

So the circular velocity of the space shuttle is 7.72 km/s. Multiply by 60 seconds per minute and by 60 minutes per hour to find that this is nearly 28,000 km/h.

CHAPTER 1 Physics Facts

1.6. What was the minimum speed required for Apollo 11 to leave the Earth?

The minimum speed to leave the surface is given by the escape velocity. For Apollo 11 to leave the Earth, it must have been traveling at least

$$v_e = \sqrt{\frac{2GM_{Earth}}{d}}$$

$$v_e = \sqrt{\frac{2 \cdot 6.67 \times 10^{-11} \, m^3/kg/s^2 \cdot 5.97 \times 10^{24} \, kg}{6.378 \times 10^6 \, m}}$$

$$v_e = \sqrt{1.25 \times 10^8 \, \frac{m^3 \cdot kg}{m \cdot kg \cdot s^2}}$$

$$v_e = \sqrt{1.25 \times 10^8 \, \frac{m^2}{s^2}}$$

$$v_e = 1.12 \times 10^4 \, m/s$$

$$v_e = 11.2 \, km/s$$

This may not seem very fast, if you are not used to thinking in km/s. Convert it to miles per hour by multiplying by 0.6214 miles/km, and multiplying by 3,600 seconds/hour. Now you see that the astronauts were traveling at 24,000 miles/hour.

1.7. What is the density of the Earth? How does this compare to the density of rocks (between 2,000 and 3,500 kg/m^3)? What does this mean?

The density is the mass divided by the volume. If we assume the Earth is spherical, the calculation is simplified.

$$\rho = \frac{M}{V}$$

$$\rho = \frac{M_{Earth}}{\frac{4}{3}\pi \cdot r_{Earth}^3}$$

$$\rho = \frac{5.97 \times 10^{24} \, kg}{\frac{4}{3}\pi \cdot (6.378 \times 10^6 \, m)^3}$$

$$\rho = 5,500 \, kg/m^3$$

The average density of the Earth is higher than the density of rock. Since the surface of the Earth is mostly rock, or water, which is even less dense, this means that the core must be made of material that is denser than the surface.

1.8. There are about 7,000 asteroids in our solar system. Assume each one has a mass of 10^{17} kg. What is the total mass of all the asteroids? If these asteroids are all rocky, and so have a density of about 3,000 kg/m^3, how large a planet could be formed from them?

The total mass of all the asteroids is just the product of the number of asteroids and their individual mass:

$$M = n \cdot m$$

$$M = 7,000 \cdot 10^{17} \, kg$$

$$M = 7 \times 10^{20} \, kg$$

The volume of the planet that could be formed is

CHAPTER 1 Physics Facts

$$V = M/\rho$$
$$V = \frac{7 \times 10^{20}\,\text{kg}}{3{,}000\,\text{kg/m}^3}$$
$$V = 2.33 \times 10^{17}\,\text{m}^3$$

If we assume the planet is spherical, then we can find the radius

$$R = \sqrt[3]{\frac{3V}{4\pi}}$$
$$R = \sqrt[3]{\frac{3 \cdot 2.33 \times 10^{17}\,\text{m}^3}{4\pi}}$$
$$R = 380\,\text{km}$$

This is a factor of about 20 less than the radius of the Earth, and about a factor of 10 less than the radius of Mars.

1.9. An asteroid's closest approach to the Sun (perihelion) is 2 AU, and farthest distance from the Sun (aphelion) is 4 AU. What is the semi-major axis of its orbit? What is the period of the asteroid? What is the eccentricity?

Figure 1-1 shows that the major axis of an orbit is the aphelion distance plus the perihelion distance. So the major axis is 6 AU, and the semi-major axis is 3 AU. The period, then, can be found from

$$P^2 = a^3$$
$$P = \sqrt{3^3}$$
$$P = \sqrt{27}$$
$$P = 5.2\,\text{years}$$

The period of the asteroid is a little over 5 years.

$$FF' = \text{aphelion} - \text{perihelion} = 2\,\text{AU}$$

$$e = \frac{FF'}{2a} = \frac{2}{6} \approx 0.33$$

The eccentricity of the elliptical orbit is 0.33.

1.10. Halley's comet has an orbital period of 76 years, and its furthest distance from the Sun is 35.3 AU. How close does Halley's comet come to the Sun? How does this compare to the Earth's distance from the Sun? What is the orbit's eccentricity?

Since Halley's comet orbits the Sun, we can use the simplified relation

$$P^2 = a^3$$

$$P^2 = \left(\frac{\text{perihelion} + \text{aphelion}}{2}\right)^3$$

$$\text{perihelion} + \text{aphelion} = 2 \cdot \sqrt[3]{P^2}$$

$$\text{perihelion} = 2 \cdot \sqrt[3]{P^2} - \text{aphelion}$$

$$\text{perihelion} = 2 \cdot \sqrt[3]{76^2} - 35.3$$

$$\text{perihelion} = 2 \cdot \sqrt[3]{5{,}776} - 35.3$$

$$\text{perihelion} = 35.8 - 35.3$$

$$\text{perihelion} = 0.5\,\text{AU}$$

The distance of closest approach of Halley's comet to the Sun is 0.5 AU. This is closer than the average distance between the Earth and the Sun.

$$e = \frac{FF'}{2a} = \frac{\text{aphelion} - \text{perihelion}}{2a}$$

$$e = \frac{34.8}{35.8} = 0.97$$

The eccentricity of this comet's orbit is very high: 0.97.

1.11. How would the gravitational force between two bodies change if the product of their masses increased by a factor of four?

The easiest way to do this problem is to begin by setting up a ratio. Since the radii stay constant, lots of terms will cancel out (see Problem 1.4):

$$\frac{F_2}{F_1} = \frac{\frac{G(mM)_2}{r^2}}{\frac{G(mM)_1}{r^2}}$$

$$\frac{F_2}{F_1} = \frac{(mM)_2}{(mM)_1}$$

$$\frac{F_2}{F_1} = \frac{4(mM)_1}{(mM)_1}$$

$$\frac{F_2}{F_1} = 4$$

The force between the two objects increases by a factor of four when the product of the masses increases by a factor of four.

1.12. How would the gravitational force between two bodies change if the distance between them increased by a factor of two?

Again, set up a ratio so that all the unchanged quantities cancel out (as in Problem 1.4):

CHAPTER 1 Physics Facts

$$\frac{F_2}{F_1} = \frac{\frac{GmM}{r_2^2}}{\frac{GmM}{r_1^2}}$$

$$\frac{F_2}{F_1} = \frac{r_1^2}{r_2^2}$$

$$\frac{F_2}{F_1} = \frac{r_1^2}{(2r_1)^2}$$

$$\frac{F_2}{F_1} = \frac{r_1^2}{4r_1^2}$$

$$\frac{F_2}{F_1} = \frac{1}{4}$$

The force between the two objects would decrease by a factor of four when the distance between them decreases by a factor of two.

1.13. How would the gravitational force between two bodies change if their masses increase by a factor of four, **and** the distance between them increased by a factor of two?

Since increasing the masses by a factor of four *increases* the force by a factor of four (Problem 1.11), and increasing the distance between them by a factor of two *decreases* the force by a factor of four (Problem 1.12), the two effects cancel out, and there is no change in the force.

1.14. What is the mass of the Sun?

Since we know the orbital period of the Earth (1 year = 3.16×10^7 seconds), and we know the orbital radius of the Earth (1 AU = 1.5×10^{11} m), we have enough information to calculate the mass of the Sun:

$$P^2 = \frac{4\pi^2 a^3}{G(m+M)}$$

$$(m+M) = \frac{4\pi^2 a^3}{GP^2}$$

Assume the mass of the Earth is small compared with the mass of the Sun ($m + M \approx M$):

$$M = \frac{4\pi^2 (1.5 \times 10^{11} \text{ m})^3}{(6.67 \times 10^{-11} \text{ m}^3/\text{kg/s}^2)(3.16 \times 10^7 \text{ s})^2}$$

$$M = 2.0 \times 10^{30} \text{ kg}$$

This is strikingly close to the accepted value for the mass of the Sun, 1.9891×10^{30} kg. It is so close that any differences might be caused by a round-off error in our calculators plus the assumption that the mass of the Earth is negligible.

1.15. How fast would a spacecraft in solar orbit have to be moving at the distance of Neptune to leave the solar system?

The escape velocity is given by

$$v_e = \sqrt{\frac{2GM}{d}}$$

$$v_e = \sqrt{\frac{2 \cdot 6.67 \times 10^{-11}\,\text{m}^3/\text{kg/s}^2 \cdot 2 \times 10^{30}\,\text{kg}}{4.5 \times 10^{12}\,\text{m}}}$$

$$v_e = 7{,}700\,\text{m/s} = 7.7\,\text{km/s}$$

In order for a spacecraft to escape the solar system from the orbit of Neptune, it must be traveling at least 7.7 km/s. This is not very much less than the escape velocity of a spacecraft from the Earth (11 km/s). Even though the orbit of Neptune is so far away, the mass of the Sun is so large that objects are bound quite tightly to the solar system, and must be moving very quickly to escape.

1.16. The Moon orbits the Earth once every 27.3 days (on average). How far away is the Moon from the Earth?

We cannot use the simple relation between P and a for this problem, since the Sun is not at the focus of the orbit. However, we **can** assume that the Moon is much less massive than the Earth. First, convert 27.3 days to 2.36×10^6 seconds.

$$a^3 = \frac{P^2 \cdot G \cdot (m+M)}{4\pi^2}$$

$$a^3 = \frac{(2.36 \times 10^6\,\text{s})^2 \, 6.67 \times 10^{-11}\,\text{m}^3/\text{kg/s}^2 (6 \times 10^{24}\,\text{kg})}{4 \cdot \pi^2}$$

$$a^3 = 5.64 \times 10^{25}\,\text{m}^3$$

$$a = 384{,}000{,}000\,\text{m}$$

$$a = 3.84 \times 10^8\,\text{m}$$

Again, this is strikingly close to the generally accepted value for the distance of the Moon (3.844×10^8 m).

1.17. What happens to the orbital period of a binary star system (a pair of stars orbiting each other) when the distance between the two stars doubles?

This orbit question requires the same ratio method as was used in Problem 1.11, but this time we need to use the equation relating P^2 and a^3:

$$\frac{P_2^2}{P_1^2} = \frac{\frac{4\pi a_2^3}{G(m+M)}}{\frac{4\pi a_1^3}{G(m+M)}}$$

$$\frac{P_2^2}{P_1^2} = \frac{a_2^3}{a_1^3}$$

$$\frac{P_2^2}{P_1^2} = \frac{(2a_1)^3}{a_1^3}$$

$$\frac{P_2^2}{P_1^2} = \frac{8a_1^3}{a_1^3}$$

$$\frac{P_2}{P_1} = \sqrt{8}$$

$$\frac{P_2}{P_1} = 2.8$$

$$P_2 = 2.8 \cdot P_1$$

The period increases by a factor of 2.8 when the distance between the two stars doubles.

CHAPTER 1 Physics Facts

About Gases

Gases made up of atoms or molecules, like the atmosphere of the Earth, are called **neutral gases**. When the atoms and molecules are ionized, so that there are electrons and ions (positively charged particles) roaming freely, the gas is called **plasma**. Plasmas have special properties, because they interact with the magnetic field. Neutral gases will not, in general, interact with the magnetic field.

THE IDEAL GAS LAW

Gases that obey the **ideal gas law** are called ideal gases. The ideal gas law states:

$$PV = NkT$$

where P is the pressure, V is the volume, N is the number of particles, T is the absolute temperature, and k is Boltzmann's constant $(1.38 \times 10^{-23}$ J/K). Sometimes both sides of this equation are divided by V, to give

$$P = nkT$$

where n is the number density (number of particles per m^3). The absolute temperature T is obtained by adding 273 to the temperature in the Celsius scale, and is measured in degrees kelvin, K. For example, 25°C is equal to 298 K. The ideal gas law provides a simple qualitative description of real gases. For example, it shows that when the volume is held constant, increasing the temperature increases the pressure. The ideal gas law is a good description of the behavior of normal stars, but fails completely for objects such as neutron stars where the gas is degenerate, and the pressure and the temperature are no longer related to each other in this way.

AVERAGE SPEED OF PARTICLES IN A GAS

The particles in a gas are moving in random directions, with speeds that depend on the temperature. Hotter gases have faster particles, and cooler gases have slower particles, on average. The average speed depends on the mass, m, of the particles:

$$v = \sqrt{\frac{8kT}{\pi m}}$$

This equation gives the *average* speed of the particles in a gas. There will be some particles moving faster than this speed, and some moving slower. If this average speed is greater than 1/6 the escape velocity of a planet, the gas will eventually escape, and the planet will no longer have an atmosphere.

This equation only holds for an ideal gas under equilibrium conditions, where it is neither expanding nor contracting, for example.

CHAPTER 1 Physics Facts

 Solved Problems

1.18. The average speed of atoms in a gas is 5 km/s. How fast will they move if the temperature increases by a factor of four?

This is another problem where we need to use a ratio (such as Problem 1.11), since the mass of the atoms in the gas is not known.

$$\frac{v_2}{v_1} = \frac{\sqrt{\frac{8kT_2}{\pi \cdot m}}}{\sqrt{\frac{8kT_1}{\pi \cdot m}}}$$

$$\frac{v_2}{v_1} = \frac{\sqrt{T_2}}{\sqrt{T_1}}$$

$$\frac{v_2}{v_1} = \sqrt{\frac{T_2}{T_1}}$$

$$\frac{v_2}{5\,\text{km/s}} = \sqrt{\frac{4T_1}{T_1}}$$

$$v_2 = 5 \cdot 2\,\text{km/s}$$

$$v_2 = 10\,\text{km/s}$$

So the speed of the atoms is doubled when the temperature increases by a factor of four.

1.19. What is the average speed of nitrogen molecules ($m = 4.7 \times 10^{-26}$ kg) at 75°F?

First, convert 75°F to degrees kelvin.

$$T_{\text{Celsius}} = (T_{\text{Fahrenheit}} - 32) \cdot \frac{5}{9}$$

$$T_{\text{Celsius}} = 24$$

$$T_{\text{Kelvin}} = T_{\text{Celsius}} + 273 = 297\,\text{K}$$

Now find the velocity:

$$v = \sqrt{\frac{8kT}{\pi \cdot m}}$$

$$v = \sqrt{\frac{8(1.38 \times 10^{-23}\,\text{J/K})297\,\text{K}}{\pi \cdot 4.7 \times 10^{-26}\,\text{kg}}}$$

$$v = 470\,\text{m/s}$$

So the molecules are moving at an average speed of 470 m/s (over 1,000 miles/hour) at room temperature!

1.20. If the temperature of a gas increases by a factor of two, what happens to the pressure (assume the volume stays the same)?

CHAPTER 1. Physics Facts

The ideal gas law states that $PV = NkT$. If the temperature (on the right-hand side of the equation) is multiplied by two, then the pressure (on the left-hand side) must also be multiplied by two. So the pressure doubles.

1.21. What is a plasma? Why does this only happen at high temperatures?

Plasmas are ionized gases, where the electrons have enough energy to be separated from the nuclei of the atoms or molecules. This happens only at high temperatures, because at lower temperatures the electrons do not have enough energy to separate themselves from the nuclei.

1.22. What is the average speed of hydrogen atoms ($m = 1.67 \times 10^{-27}$ kg) in the Sun's photosphere ($T \sim 5{,}800$ K)?

$$v = \sqrt{\frac{8kT}{\pi \cdot m}}$$

$$v = \sqrt{\frac{8 \cdot 1.38 \times 10^{-23}\,\text{J/K} \cdot 5{,}800\,\text{K}}{\pi \cdot 1.67 \times 10^{-27}\,\text{kg}}}$$

$$v = 11{,}000\,\text{m/s}$$

The hydrogen atoms in the Sun's photosphere are moving at about 11,000 m/s. To convert to something (slightly) more familiar, multiply by 0.0006 to convert meters to miles, and by 3,600 to convert the seconds to hours. The hydrogen atoms are traveling nearly 24,000 miles per hour. It would take about 1 hour for one of these atoms to travel all the way around the Earth.

About Light

Light exhibits both particle behavior, giving momentum to objects it strikes, and wave behavior, bending as it crosses a boundary into a lens or a prism. Light can be described in terms of electromagnetic waves, or as particles, called "photons."

WAVELENGTH, FREQUENCY, AND SPEED

The **wavelength** of a wave of any kind is the distance between two successive peaks (see Fig. 1-2). The **frequency** is the number of waves per second that pass a given point. If you are standing on the shore, you can count up the number of waves that come in over 10 seconds, and divide that number by 10 to obtain the frequency.

Mathematically, the frequency, f, the speed, v, and the wavelength, λ, are all related to one another by the following equation:

$$v = \lambda \cdot f$$

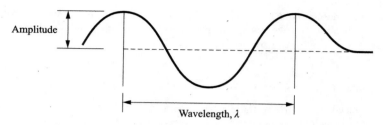

Fig. 1-2. A wave. The wavelength is the distance between crests, and the amplitude is the height of a crest.

In a vacuum, such as interstellar space, the speed of light is 3×10^8 m/s for all wavelengths. This is the speed of light, and is always designated by c.

Rearranging the above equation (and substituting c for the speed), we see that in a vacuum,

$$\lambda = \frac{c}{f}$$

Since the speed of light, c, is a constant, we can see that λ and f are inversely proportional to each other, so that if one gets larger, the other gets smaller. Therefore, long-wavelength waves have low frequencies, and short-wavelength waves have high frequencies.

VISIBLE LIGHT AND COLOR

The visible part of the whole range of wavelengths is only a small part of the entire range of light (see Fig. 1-3). The color of visible light is related to its wavelength. Long-wavelength light is redder, and short-wavelength light is bluer. Longer than red is infrared, microwave and radio, and shorter than blue is ultraviolet, X-rays, and gamma rays.

The energy of a photon is given by

$$E = hf$$

or in terms of wavelength,

$$E = hc/\lambda$$

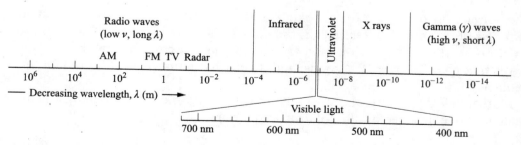

Fig. 1-3. The entire spectrum of light.

CHAPTER 1 Physics Facts

where h is Planck's constant ($h = 6.624 \times 10^{-32}$ J·s). Photons can collide with and give their energy to other particles, such as electrons. In contrast to electrons and other particles, the photon has zero rest mass and in vacuum always travels at the same speed, c.

There are only two wavelength bands where light can come through the atmosphere unobstructed: the visible and the radio. In most other wavelengths, the sky is opaque. For example, the atmosphere keeps out most of the gamma rays (high-energy light).

Each of the wavelength bands (radio, visible, gamma ray, UV, etc.) has a special kind of telescope for observing. The most familiar kinds are optical (visible) and radio telescopes. These are usually ground-based. The atmosphere is mostly opaque in the other bands such as X-ray, and telescopes observing at these wavelengths must be placed outside of the atmosphere.

SPECTRA

The emission spectrum is a graph of energy emitted by an object at each wavelength (Fig. 1-4). If you do this for multiple objects, then you have many spectra. Similarly, the amount of light absorbed would be an absorption spectrum.

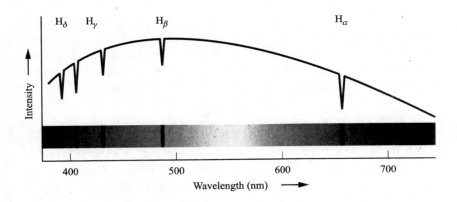

Fig. 1-4. A stellar spectrum.

BLACKBODY EMISSION

The spectrum of blackbody emission has a very special shape, as shown in Fig. 1-5. This kind of emission is also called "continuous emission," because emission occurs at all wavelengths, contrary to line emission (see below).

Both the height of the curve and the wavelength of the peak change with the temperature of the object. The three curves in the figure show what happens to the blackbody emission of an object as it is heated. When the object is cool, the

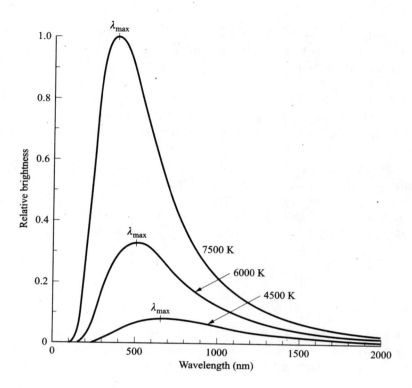

Fig. 1-5. Blackbody emission.

strongest emission is in the red, and as it gets hotter, the strongest part of the spectrum moves towards the blue. We can find the temperature, T (in kelvin), of an object by looking at its blackbody emission, and finding the wavelength of the peak, λ_{max} (in meters). Wien's Law relates these two quantities:

$$\lambda_{max} = \frac{0.0029 \, \text{m} \cdot \text{kelvin}}{T}$$

The height of the curve also changes with temperature, which tells you that the energy emitted must change (because you have more light coming from the object, and light is a form of energy). The Stefan-Boltzmann Law relates the energy emitted per second per unit area of the surface of an object, to the temperature of the object:

$$\Im = \sigma \cdot T^4$$

where σ is the Stefan-Boltzmann constant, equal to $5.6705 \times 10^{-8} \, \text{W/m}^2 \, \text{K}^4$.

The light coming from stars has the general shape of a blackbody. So does the infrared light coming from your body. Everything emits light with a spectrum of this shape, with intensity and color depending on its temperature. Most objects, including stars, also have other things going on as well, so that the spectrum is almost never a pure blackbody. For example, the object may not be evenly heated, or there may be line emission contributing (see below). We know of only one perfect blackbody, and that is the Universe itself, which has the black-

CHAPTER 1 Physics Facts

body spectrum of a body at a temperature of 2.74 K. This is called the cosmic microwave background radiation (CMBR).

LINE EMISSION

Line emission is produced by gases. The spectrum consists of bright lines at particular frequencies characteristic of the emitting atoms or molecules. Atoms consist of a nucleus (positively charged), surrounded by a cloud of electrons (negatively charged). When light strikes the electrons, it gives them extra energy. Because they have more energy, they move farther from the nucleus (out through the "valence levels" or "energy levels" or "shells"). But it is a rule in the Universe that objects prefer to be in their lowest energy states (this is why balls roll downhill, for example). So the electrons give up energy in the form of light in order to move to the lowest energy level, also known as the "ground state." It is important to note that these levels are discrete, not continuous. It's like the difference between climbing steps and climbing a ramp. There are only certain heights in a flight of steps, and similarly there are only certain amounts of energy that an electron can take or give up at any given time (Fig. 1-6). When light has been taken out of a spectrum, it is called an absorption line (see Fig. 1-7).

The energy emitted when an electron makes a transition from an energy level E_h to a lower energy level E_l is $E_h - E_l$ and is emitted in the form of a photon of frequency

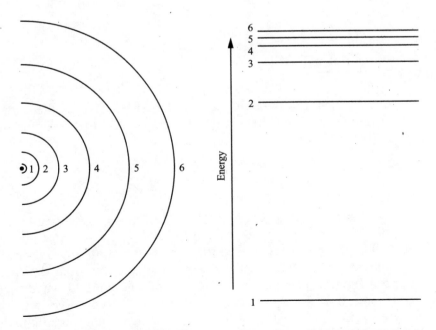

Fig. 1-6. Energy levels in an atom. When electrons move up, they **absorb** light. When they move down, they **emit** light.

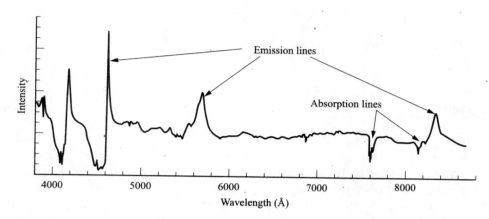

Fig. 1-7. Emission lines occur at frequencies where energy is added to a spectrum, and absorption lines occur at frequencies where energy is subtracted from a spectrum.

$$f = (E_h - E_l)/h$$

where h is Planck's constant. The reverse process, moving from a lower energy level to a higher energy level, requires the same amount of energy, and is accomplished by absorbing a photon of the same frequency, f. Thus, the absorption lines for a given gas occur at the same frequencies as the emission lines. Furthermore, as the values of the energy levels, E_h, E_l, etc., are different for each atom, the absorption/emission lines will occur at frequencies that are characteristic of each kind of atom. The absorption/emission spectrum can be used to identify various atoms.

THE DOPPLER EFFECT

When an object is approaching or moving away, the wavelength of the light it emits (or reflects) is changed. The shift of the wavelength, $\Delta\lambda$, is directly related to the velocity, v, of the object:

$$\Delta\lambda = \frac{\lambda_0 v}{c}$$

Here λ_0 is the wavelength emitted if the object is at rest and v is the component of the velocity along the "line of sight" or the "radial velocity." Motion perpendicular to the line of sight does not contribute to the shift in wavelength. This equation holds when the velocity of the object is much less than the speed of light. For approaching objects, $\Delta\lambda$ is negative, and the emitted wavelength appears shorter ("blue-shifted"). For receding objects, $\Delta\lambda$ is positive, and the emitted wavelength appears longer ("red-shifted"). This is roughly analogous to the waves produced by a boat in the water.

CHAPTER 1 Physics Facts

ALBEDO

The albedo of an object is the fraction of light that it reflects. The albedo of an object can be any value between 0 and 1, where 0 implies all the light is absorbed, and 1 implies all the light is reflected. Mirrors have high albedos, and coal has a very low albedo.

Solved Problems

1.23. A sound wave in water has a frequency of 256 Hz and a wavelength of 5.77 m. What is the speed of sound in water?

Use the relationship between wavelength, speed, and frequency,

$$s = \lambda \cdot f$$
$$s = 5.77 \cdot 256 \frac{\text{m}}{\text{s}}$$
$$s = 1,480 \, \text{m/s}$$

The speed of sound in water is 1,480 m/s.

1.24. Why don't atoms emit a continuous spectrum?

Emission from atoms is produced when the electrons drop from an initial energy level E_i to a final energy level E_f. The difference in energy, $E_f - E_i$, is given up as a photon of wavelength

$$\lambda = \frac{hc}{E_f - E_i}$$

Because these levels are quantized (step-like), the electrons can give up only certain amounts of energy each time they move from one energy level to another. Since energy is related to frequency, this means that the photons can only have certain frequencies, and therefore certain wavelengths. This is exactly what we mean when we say the emission is "line emission"—it only exists at certain wavelengths.

1.25. In what wavelength region would you look for a star being born ($T \sim 1,000 \, \text{K}$)?

The continuous spectrum of the star will have maximum emission intensity at a wavelength λ_{max} given by Wien's Law:

$$\lambda_{max} = \frac{0.0029}{T}$$
$$\lambda_{max} = \frac{0.0029}{1,000}$$
$$\lambda_{max} = 2.9 \times 10^{-6} \, m$$

From Fig. 1-3, we can see that this wavelength is in the infrared region of the spectrum. So in order to search for newly forming stars, we should observe in the infrared.

1.26. Two satellites have different albedos: one is quite high, 0.75, and the other is quite low 0.15. Which is hotter? Why are satellites usually made of (or covered with) reflective material?

The satellite with the higher albedo reflects more light, and therefore is cooler than the satellite with the lower albedo. Satellites have reflective material on the outside to keep the electronics cool on the inside.

1.27. Your body is about 300 K. What is your peak wavelength?

$$\lambda_{max} = \frac{0.0029 \, m \cdot K}{T}$$
$$\lambda_{max} = \frac{0.0029 \, m \cdot K}{300 \, K}$$
$$\lambda_{max} = 9.7 \times 10^{-6} \, m$$

Your peak wavelength is 9.7×10^{-6} m. From Fig. 1-3, you can see that this is in the infrared, which means that in a dark room (so that you are not reflecting any light) you should appear brightest through a pair of infrared goggles, which is the basis of night vision devices.

1.28. What is the energy of a typical X-ray photon?

From Fig. 1-3, we can see that the middle of the X-ray part of the spectrum is about 10^{19} Hz. To find the energy, use

$$E = h \cdot f$$
$$E = 6.624 \times 10^{-34} \, J \cdot s \cdot 10^{19} \, Hz$$
$$E = 6.624 \times 10^{-15} \, J$$

1.29. How long does it take light to reach the Earth from the Sun?

From the distance between the Earth (1.5×10^{11} km) and the Sun, and the speed of light (3×10^5 km/s), we can calculate the time it takes for light to make the trip:

$$t = \frac{d}{v}$$
$$t = \frac{1.5 \times 10^{11} \, m}{3 \times 10^8 \, m/s}$$
$$t = 500 \, s$$

Dividing this by 60 to convert to minutes gives about 8.3 minutes for light to travel from the Earth to the Sun.

CHAPTER 1 Physics Facts

1.30. What is the frequency of light with a wavelength of 18 cm? What part of the spectrum is this?

First, find the frequency.

$$\lambda = \frac{c}{f}$$

can be rearranged to give

$$f = \frac{c}{\lambda}$$
$$f = \frac{3 \times 10^8 \text{ m/s}}{0.18 \text{ m}}$$
$$f = 1.7 \times 10^9 \text{ Hz}$$

From Fig. 1-3, we can see that this is in the radio part of the spectrum. If we want to observe at this wavelength, we must use a radio telescope.

1.31. One photon has half the energy of another. How do their frequencies compare?

Because the energies are directly proportional to the frequencies,

$$E = h \cdot f$$

if the energy (on the left-hand side) is reduced by half, the right-hand side must also be reduced by half: h is a constant, so one photon must have half the frequency of the other.

1.32. How many different ways can an electron get from state 4 to the ground state (state 1) in a hydrogen atom? Sketch each one. What does this tell you about the expected spectrum of hydrogen gas?

There are four different ways that the electron can go from state 4 to state 1. From the diagram of all the different paths (Fig. 1-8), you can see that there are six distinct steps that can be taken. This means that there can be at least six different emission lines produced by a group of atoms going from state 4 to state 1. For the hydrogen atom,

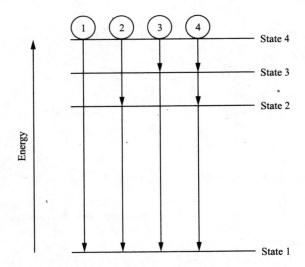

Fig. 1-8. There are four different ways for the electron to move from state 4 to state 1.

electron transitions ending at the ground state (state 1) produce a series of spectral lines known as the Lyman series.

1.33. Suppose molecules at rest emit with a wavelength of 18 cm. You observe them at a wavelength of 18.001 cm. How fast is the object moving and in which direction, towards or away from you?

This problem calls for the Doppler equation. λ_0 is 18 cm, $\Delta\lambda$ is $(18.001 - 18) = 0.001$ cm.

$$\frac{v}{c} = \frac{\Delta\lambda}{\lambda_0}$$
$$v = \frac{0.001}{18} 3 \times 10^8 \text{ m/s}$$
$$v \cong 17{,}000 \text{ m/s}$$

The molecules are traveling with a radial speed of 17 km/s. Since the wavelength is longer, the light has been red-shifted, or stretched out, and so the object is moving away from you. The object may also be moving across your field of view, but this motion will not contribute to the Doppler shift.

1.34. How much energy is radiated into space by each square meter of the Sun every second ($T \sim 5{,}800$ K)? What is the total power output of the Sun?

Use the Stefan-Boltzmann law:

$$\Im = \sigma \cdot T^4$$
$$\Im = 5.67 \times 10^{-8} \text{ W/m}^2/\text{K}^4 \cdot (5{,}800)^4$$
$$\Im = 6.4 \times 10^7 \text{ W/m}^2$$

To find the total power output of the Sun, we must multiply by the surface area of the Sun. First, find the surface area,

$$A = 4\pi r^2$$
$$A = 4\pi \cdot (7 \times 10^8 \text{ m})^2$$
$$A = 6.16 \times 10^{18} \text{ m}^2$$

Now calculate the power by multiplying $\Im$ by A to get 3.9×10^{26} watts. This is quite close to the accepted value of 3.82×10^{26} watts. The discrepancy probably comes from round-off error in the surface temperature, in the constants, and in the calculation.

1.35. What information can you find out from an object's spectrum?

The temperature of an object can be determined from the peak wavelength of the spectrum. The composition of the object can be determined from the presence or absence of emission lines of particular elements, the speed at which the object is approaching or receding can be determined from the Doppler shift, and if there are absorption lines present, you know that there is cool gas between you and the object you are observing. All together, a spectrum of a star or other astronomical object contains a great deal of information.

About Distance

THE SMALL ANGLE FORMULA

The small angle formula (Fig. 1-9) indicates how the size of an object appears to change with distance. We measure the apparent size of an object by measuring the angle from one side to the other. For example, the Moon is about 0.5° across. So is the Sun, even though it is actually **much** larger than the Moon. The relationship between the angular diameter, θ, the actual diameter, D, and the distance, d, is

$$\theta('') = 206{,}265 \cdot \frac{D}{d}$$

where θ is measured in **arcseconds** (''). An arcsecond is 1/60 of an **arcminute** ('), which is 1/60 of a degree (°). The angular diameter of a tennis ball, 8 miles away, is 1 arcsecond.

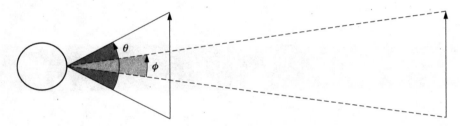

Fig. 1-9. The small angle relation. The farther away an object is, the smaller it appears to be.

THE INVERSE SQUARE LAW

Assume that an object, for example a star, is uniformly radiating energy in all directions at a rate of E watts per second. Consider a large spherical surface of radius d, centered on the star. The amount of energy received per unit area of the sphere per second is

$$F = \frac{E}{4\pi d^2}$$

Thus, the amount of starlight reaching a telescope is inversely proportional to the square of the distance to the star (Fig. 1-10). For example, if two stars A and B are identical, and star B is twice as far from Earth as star A, then star B will appear four times dimmer.

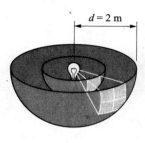
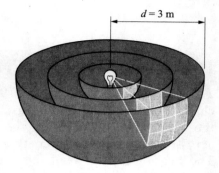

Flux $(F) = \dfrac{\text{Energy}}{\text{Area}} = \dfrac{E}{4\pi d^2}$

Distance $(d) = 1$ meter, so

$F = \dfrac{E}{4\pi 1^2} = \dfrac{E}{4\pi}$

Distance $(d) = 2$ meters, so

Flux $(F) = \dfrac{E}{4\pi d^2} = \dfrac{E}{4\pi 2^2} = \dfrac{E}{16\pi}$

Notice: Flux (F) decreases as distance (d) increases because energy (E) remains the same but area of sphere $(4\pi d^2)$ increases

Distance $(d) = 3$ meters, so

Flux $(F) = \dfrac{E}{4\pi 3^2} = \dfrac{E}{36\pi}$

Fig. 1-10. The inverse square law. As your distance to a shining object increases, you receive less light.

Solved Problems

1.36. Supernova remnants expand at about 1,000 km/s. Given a remnant that is 10,000 pc away, what is the change in angular diameter over 1 year (pc = parsec, a unit of distance, see Appendix 1)?

First, find the linear expansion over 1 year, by multiplying the speed by the amount of time:

$$D = v \cdot t$$
$$D = (1{,}000 \text{ km/s}) \cdot (3.16 \times 10^7 \text{ s})$$
$$D = 3.2 \times 10^{10} \text{ km}$$

Now, use the small angle formula to find the corresponding angular size:

$$\theta('') = 206{,}265 \cdot \dfrac{D}{d}$$
$$\theta('') = 206{,}265 \cdot \dfrac{3.2 \times 10^{10} \text{ km}}{10{,}000 \text{ pc}}$$
$$(1 \text{ pc} = 3.1 \times 10^{13} \text{ km})$$
$$\theta('') = 206{,}265 \cdot \dfrac{3.2 \times 10^{10} \text{ km}}{10{,}000 \cdot 3.1 \times 10^{13} \text{ km}}$$
$$\theta('') = 0.02''$$

So even such a large expansion, when so far away, is barely observable. This angular expansion is just 1/50 the angular size of a tennis ball 8 miles away.

CHAPTER 1 Physics Facts

1.37. The Moon and the Sun are about the same size in the sky (0.5°). Given that the diameter of the Moon is about 3,500 km, and the diameter of the Sun is about 1,400,000 km, how much farther away is the Sun than the Moon?

Use the ratio method.

$$\frac{\theta_{Sun}}{\theta_{Moon}} = \frac{206{,}265 \frac{D_{Sun}}{d_{Sun}}}{206{,}265 \frac{D_{Moon}}{d_{Moon}}}$$

$$\frac{d_{Sun}}{d_{Moon}} = \frac{D_{Sun}}{D_{Moon}}$$

$$\frac{d_{Sun}}{d_{Moon}} = \frac{1{,}400{,}000 \text{ km}}{3{,}500 \text{ km}}$$

$$\frac{d_{Sun}}{d_{Moon}} = 400$$

So the distance from the Earth to the Sun is about 400 times larger than the distance from the Earth to the Moon!

1.38. Europa (a moon of Jupiter) is five times further than the Earth from the Sun. What is the ratio of flux at Europa to the flux at the Earth?

This is another problem that is easiest when solved as a ratio. In fact, since a ratio is what you are looking for, it's especially well suited to this method.

$$\frac{F_{Europa}}{F_{Earth}} = \frac{\frac{E_{total}}{4\pi \cdot d^2_{Europa-Sun}}}{\frac{E_{total}}{4\pi \cdot d^2_{Earth-Sun}}}$$

$$\frac{F_{Europa}}{F_{Earth}} = \frac{d^2_{Earth-Sun}}{d^2_{Europa-Sun}}$$

But, $d_{Europa-Sun} = 5 d_{Earth-Sun}$, so

$$\frac{F_{Europa}}{F_{Earth}} = \frac{d^2_{Earth-Sun}}{(5 d^2_{Europa-Sun})^2}$$

$$\frac{F_{Europa}}{F_{Earth}} = \frac{1}{25}$$

So Europa receives 25 times less sunlight than the Earth does.

Supplementary Problems

1.39. What is the force of gravity between the Earth and the Sun?

Ans. 3.52×10^{22} N

1.40. If someone weighs 120 pounds on Earth, how much do they weigh on the Moon?

Ans. 20 pounds

1.41. What is the circular velocity of the Moon?

Ans. 1,000 m/s

1.42. What is the density of the Moon? How does this compare to the density of rocks?

Ans. 3,300 kg/m^3, about the same as rock

1.43. An asteroid's semimajor axis is 3.5 AU. What is its period?

Ans. 6.5 years

1.44. How would the gravitational force between two bodies change if the distance between them decreased by a factor of four?

Ans. Force decreases by a factor of 16

1.45. What is the minimum speed of the solar wind as it leaves the photosphere? (What is the escape velocity for particles leaving the surface of the Sun?)

Ans. 619,000 m/s

1.46. Mars orbits the Sun once every 1.88 years. How far is Mars from the Sun?

Ans. 1.52 AU

1.47. Phobos orbits Mars once every 0.32 days, at a distance of 94,000 km. What is the mass of Mars?

Ans. 6.43×10^{23} kg

1.48. The speed of atoms in a gas is 10 km/s. How fast will they move if the temperature decreases by a factor of two?

Ans. 7.07 km/s

1.49. What is the speed of electrons ($m = 9.1094 \times 10^{-34}$ kg) in the Sun's photosphere (5,800 K)?

Ans. 1.5×10^4 km/s

CHAPTER 1 Physics Facts

1.50. If the pressure of a gas increases by a factor of four, and the temperature stays the same, what happens to the volume?

Ans. Volume decreases by a factor of four

1.51. An object has an albedo of 0.7 and receives a flux of 100 W/m². What is the reflected flux of the object?

Ans. 70 W/m²

1.52. What is the maximum wavelength of radiation emitted from your desk (room temperature $= 297\,\text{K}$)?

Ans. $9.7 \times 10^{-6}\,\text{m}$

1.53. What is the energy of a typical photon of visible light ($\lambda = 5 \times 10^{-7}\,\text{m}$)?

Ans. $4 \times 10^{-19}\,\text{J}$

1.54. How long does it take a signal to reach Neptune from the Earth (calculate at the closest distance between them)?

Ans. 4 hours

1.55. A photon with rest wavelength 21 cm is emitted from an object traveling towards you at a velocity of 100 km/s. How does its energy compare to the rest energy?

Ans. The ratio of energies is 1.00079

1.56. How much energy is radiated from each square meter of the surface of the Earth every second (assume $T = 300\,\text{K}$)?

Ans. 459 W/m²

1.57. What is the angular size of Jupiter from the Earth (at the closest distance between them)?

Ans. 47″

1.58. How close would the Earth have to be to the Sun for the gravitational force between the Sun and the Earth to be equal to the gravitational force between the Moon and the Earth?

Ans. $2 \times 10^{12}\,\text{m}$

CHAPTER 2

The Sky and Telescopes

Coordinate Systems and Timescales

COORDINATE SYSTEMS

Altitude and Azimuth. The **altitude** is defined relative to the **horizon**, and is the angle from the horizon to the object. The altitude of an object on the horizon is 0°, and the altitude of an object directly overhead is 90°. The point directly overhead (altitude = 90°) is the **zenith**, and a line running from north to south through the zenith is called the local **meridian**. **Azimuth** is the angle around the horizon from north and towards the east. An object in the north has azimuth 0°, while an object in the west has azimuth 270°. The altitude and azimuth of an object are particular to the observing location and the time of observation.

Right ascension and declination. **Declination** (dec) is like latitude on the Earth, and measures the angle north and south of the **celestial equator** (an imaginary line in the sky directly over the Earth's equator). The celestial equator lies at 0°, while the **north celestial pole** (i.e., the extension of the Earth's rotation axis) is at 90°. Declination is negative for objects in the southern celestial hemisphere, and is equal to −90° at the south celestial pole.

Right ascension (RA) is analogous to longitude. The **ecliptic** is the plane of the solar system, or the path that the Sun follows in the sky. Because the axis of the Earth is tilted, the ecliptic and the celestial equator are not in the same place, but cross at two locations, called the equinoxes. One of these locations, the **vernal equinox**, is used as the zero point of right ascension. Right ascension is measured in hours, minutes, and seconds to the east of the vernal equinox. There are 24 hours of right ascension in the sky, and during the 24 hours of the Earth's day, all of them can be seen. Each hour on Earth changes the right ascension of the meridian by just under 1 hour.

Figure 2-1 shows a diagram of the important points in the sky. Imagine that you "unrolled" the sky and made a map, much like a map of the Earth. All of the following are labeled on the map, and many are explained in more detail later in the chapter:

(a) ecliptic
(b) celestial equator, 0 degrees dec
(c) autumnal equinox, 12 hours RA
(d) vernal equinox, 0 hours RA
(e) summer solstice, 6 hours RA
(f) winter solstice, 18 hours RA
(g) direction of the motion of the Sun throughout the year.

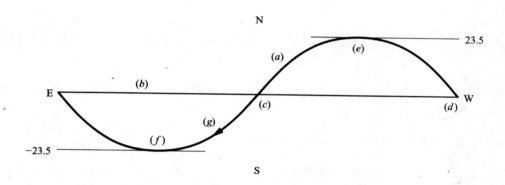

Fig. 2-1. A diagram of the sky. The path of the Sun is indicated by the curved line.

THE DAY

The Earth rotates on its axis once per day, giving us day and night. Astronomers define three different kinds of "day," depending upon the frame of reference:

1. **Sidereal day.** The length of time that it takes for the Earth to come around to the same position relative to the distant stars. The sidereal day is 23 hours and 56 minutes long. Specifically, it is measured as the time between successive meridian crossings of the vernal equinox.

2. **Solar day.** The length of time that it takes for the Earth to come around to the same position relative to the Sun. It is measured as the time between successive meridian crossings of the Sun. The solar day is 24 hours long. The "extra" four minutes come from the fact that the Earth travels about 1 degree around the Sun per day, so that the Earth has to turn a little bit further to present the same face to the Sun.

CHAPTER 2   The Sky and Telescopes

3. **Lunar day.** The length of time that it takes for the Earth to come around to the same position relative to the Moon. Since the Moon revolves around the Earth, this day is even longer than the solar day—about 24 hours and 48 minutes. This is why the tides do not occur at the same time every day, because the Moon is the primary contributor to the tides, and it is not in the same location in the sky each day.

TIDES

The Earth experiences one full set of tides each day (two highs and two lows), everywhere on the planet. Tides are caused by gravity. The Sun and the Moon both contribute to tides on Earth.

When the Sun, the Moon, and the Earth are all in a straight line, the tides are largest. This is called a spring tide (spring for jumping, not spring for the season). When the Sun, the Moon, and the Earth are at right angles, the tides are smallest. This is called a neap tide.

Tides are slowing the Earth's rotation. The rotation is slowed by about 0.0015 seconds every century. Eventually, the Moon and the Earth will become "tidally locked," so that the same face of the Earth always faces the same face of the Moon. An Earth day will slow to be about 47 of our current days long.

As the Earth is slowing down, its angular momentum decreases. Conservation of angular momentum requires that the Moon's angular momentum should increase. Indeed, the Moon increases its angular momentum by receding from the Earth (about 3 cm per year). As it gets further away, its angular size decreases, and it looks smaller relative to the Sun. It will take several centuries for this to add up to a noticeable effect.

THE MOON ORBITS THE EARTH

The Moon goes through one cycle of phases in about 29 days. This is not the same as the length of a calendar month. This is why the moon is not always new on the first day of the calendar month.

The Moon is in "synchronous rotation." The length of time that it takes to rotate on its axis is equal to the length of time it takes to revolve around the Earth. As a result, the same side of the Moon always faces the Earth.

As the moon orbits, it exhibits phases. Figure 2-2 shows the Moon at various points in its orbit around the Earth. The lighter shading indicates the illuminated portion of the Moon. This diagram is drawn from the point of view of looking down from north.

THE EARTH ORBITS THE SUN

A year is defined as the amount of time that it takes for the Earth to complete a full orbit around the Sun. A year is about 365.25 days long. This has many visible consequences.

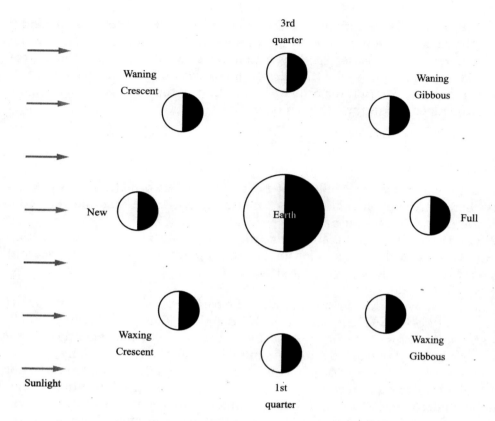

Fig. 2-2. The phases of the Moon. Depending on the relative orientation of the Earth, the Sun, and the Moon, different parts of the Moon appear lit.

1. **Visible constellations cycle.** Because the Earth is traveling around the Sun, the constellations (made up of stars that are much further away than the Sun) that are visible at night vary during the course of the year. That is, the direction that you are looking into the night sky changes. The stars rise 4 minutes earlier each night (a total of about 2 hours per month). For example, if a star rises at 6 p.m. on January 1, it will rise at 4 p.m. on February 1.

2. **Seasons.** The Earth's equator is tilted by 23.5 degrees to the plane of the Earth's orbit (the ecliptic). Because of this, the Earth has four seasons each year. When the North Pole is tipped in the direction of the Sun, it is summertime in the Northern Hemisphere, but wintertime in the Southern Hemisphere. Conversely, when the South Pole is tilted towards the Sun, it is summer in the Southern Hemisphere, and winter in the North.

   The ecliptic and the celestial equator intersect at two points, called the **equinoxes**. When the Sun has reached those points on the ecliptic, it is directly over the Earth's equator, and the whole planet gets 12 hours of sunlight and 12 hours of darkness. This happens once in the spring (vernal equinox, March 21), and once in the fall (autumnal equinox, September 21). **Solstices** occur when the Earth's North Pole is tilted farthest towards

CHAPTER 2 The Sky and Telescopes

or away from the Sun. Summer solstice (June 21) is the longest day of the year in the Northern Hemisphere, and winter solstice (December 21) is the shortest day of the year in the Northern Hemisphere.

3. **Relative motion of planets**. The Earth and the other planets change their relative positions, so that the planets appear to move generally eastward with respect to the stars and constellations over the course of the year. The Earth completes an orbit considerably faster than most of the outer planets, so that they tend to be visible at nearly the same time of year for many years in a row.

ECLIPSES

In a **solar eclipse**, the Moon comes between the Earth and the Sun, and casts its shadow on the Earth. In a **lunar eclipse**, the Earth comes between the Sun and the Moon, and the Earth's shadow is cast on the Moon. Lunar eclipses are quite common, but solar eclipses are relatively rare at any given location. A given eclipse is visible from particular geographic locations, and may not produce any effect elsewhere. Eclipses do not occur every month because the Moon's orbit around the Earth is not aligned with the Earth's orbit around the Sun (the ecliptic). The two orbits are inclined about 5 degrees with respect to each other.

PRECESSION

Precession is the change in the direction of the Earth's spin axis. Precession is easily observed in tops. A spinning top not only rotates around on its own axis, but also the axis wobbles, and points in different directions due to gravity. The axis of the Earth behaves in a similar manner due to the Sun's gravitational force, so that the North Pole points at different stars at different times. Therefore, Polaris has not always been the North Star. About 3,000 years ago, Thuban (a star in the constellation Draco) was the North Star; about 12,000 years from now, Vega (a star in the constellation Lyra) will be the North Star. It takes 26,000 years for the pole to completely precess.

As the axis points in different directions, the plane of the celestial equator moves, so that the equinoxes also move. This is often called "precession of the equinoxes." Every 50 years or so, astronomers have to recalculate the positions of all of the stars in the sky, since the origin of the declination–right ascension coordinate system (the vernal equinox) moves.

Solved Problems

2.1. What is the ratio of the gravitational force of the Sun on the Earth to the gravitational force of the Moon on the Earth?

As in Chapter 1, we should set up a ratio,

$$\frac{F_{Moon}}{F_{Sun}} = \frac{\dfrac{Gm_{Earth}M_{Moon}}{d^2_{Earth-Moon}}}{\dfrac{Gm_{Earth}M_{Sun}}{d^2_{Earth-Sun}}}$$

$$\frac{F_{Moon}}{F_{Sun}} = \frac{M_{Moon} \cdot d^2_{Earth-Sun}}{M_{Sun} \cdot d^2_{Earth-Moon}}$$

From Appendix 2,

$$\frac{F_{Moon}}{F_{Sun}} = \frac{7.35 \times 10^{22} \text{ kg} \cdot (1.5 \times 10^{11} \text{ m})^2}{2 \times 10^{30} \text{ kg} \cdot (3.84 \times 10^8 \text{ m})^2}$$

$$\frac{F_{Moon}}{F_{Sun}} = 0.0056$$

The gravitational force between the Sun and the Earth is much larger than the gravitational force between the Moon and the Earth, even though the Sun is much farther away. In this case, the Sun is so much larger than the Moon that even comparing the squares of the distances does not equalize the problem.

2.2. If the Earth rotated in the opposite sense (clockwise rather than counterclockwise), how long would the solar day be?

To complete a solar day, the Earth has to move an extra degree, or 4 minutes, than to complete the sidereal day. If it rotated in the opposite direction, this effect would be reversed. The solar day would be 1 degree, or 4 minutes, *shorter* than the sidereal day. So it would be 8 minutes shorter than the current day, or 23 hours and 52 minutes long.

2.3. If you look overhead at 6 p.m. and notice that the moon is directly overhead, what phase is it in?

At 6 p.m., the Sun is just setting, so if you observed the solar system from north, you'd see the view shown in Fig. 2-3. If the Moon were just overhead, you'd see the view shown in Fig. 2-4. The Moon is halfway in its orbit between new and full phases, so it is a quarter moon. Comparing to Fig. 2-2, you can see that the Moon is in its first quarter.

CHAPTER 2 The Sky and Telescopes

Fig. 2-3. Diagram of setting Sun, as viewed from north of the Earth.

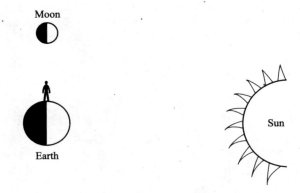

Fig. 2-4. Diagram of setting Sun, and the Moon overhead for observer on the Earth.

2.4. Which declinations can be observed from the North Pole? the South Pole? the equator? Which right ascensions can be seen from the above locations each day?

From the North Pole, declinations 0–90° north can be seen. From the South Pole, declinations 0–90° south can be seen. From the equator, all declinations can be seen.

Since the Earth rotates completely around its axis in one day, all right ascensions can be seen from every location on the Earth every day. This does not mean that all stars are seen from all locations every day—some of them are on the sunward side of the sky, and some are always below the horizon for a given location.

2.5. What is the latitude of the North Pole? Why is it impossible to give the longitude? What are the coordinates (in RA and dec) of the north celestial pole?

The North Pole lies at 90° north latitude. All longitude lines meet at the poles, and so it is impossible to determine the longitude of the North Pole.

Similarly, the north celestial pole is located at 90° north declination, and the RA cannot be determined (although it is usually denoted as 00°00m00s, by convention).

2.6. Suppose that the Earth's pole was perpendicular to its orbit. How would the azimuth of sunrise vary throughout the year? How would the length of day and night vary throughout the year at the equator? at the North and South Poles? where you live?

If the ecliptic and the celestial equator were not tilted with respect to each other, every day would be an equinox, because the Sun would always be on the celestial equator. The Sun would rise at 90° azimuth (due east) every day, and each day would be divided into 12 hours of daylight and 12 hours of darkness at all of these locations.

2.7. You are an astronaut on the moon. You look up, and see the Earth in its full phase and on the meridian. What lunar phase do people on Earth observe? What if you saw a first quarter Earth? new Earth? third quarter Earth? Draw a picture showing the geometry.

When you observe the Earth in its full phase, the Moon, the Earth, and the Sun are in a line, with the Moon between the Earth and the Sun. This means that the Moon is new. First quarter Earth corresponds to third quarter Moon, new Earth to full Moon, and third quarter Earth to first quarter Moon (Fig. 2-5). The phases are opposite.

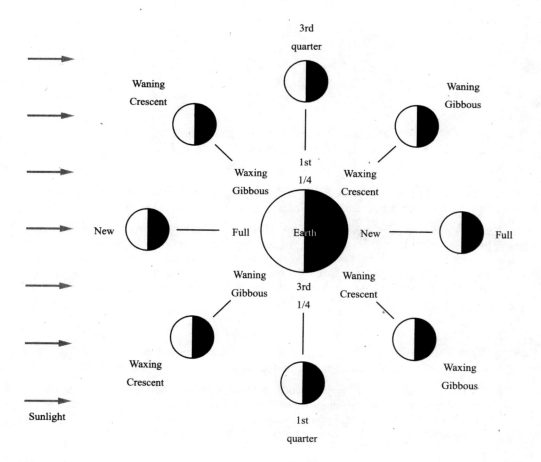

Fig. 2-5. The phases of the Earth as seen from the Moon.

CHAPTER 2 The Sky and Telescopes

2.8. Explain why some stars in your sky never rise, while others never set—assume that you live in a northern latitude (between 0° and 90° north).

We are not located at the equator, and so from where we are, the horizon is not parallel to the polar axis. From the geometry of Fig. 2-6, stars that have declinations greater than those equal to the observer's latitude will never set. These are called circumpolar stars. Stars with declinations less than the negative of the observer's latitude will never rise. Stars in between these two declinations rise and set.

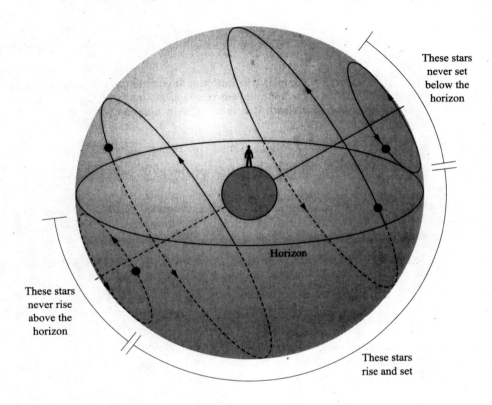

Fig. 2-6. Position of horizon and fraction of circumpolar stars.

2.9. (a) How does the declination of the Sun vary over the year? (b) Does its right ascension increase or decrease from day to day?

The declination of the Sun increases between winter and summer solstice, and decreases from summer to winter solstice. The right ascension increases from day to day, as the Sun moves eastward across the constellations. The Sun has to move from vernal equinox to summer solstice to autumnal equinox, etc., all east of each other.

2.10. If a planet always keeps the same side towards the Sun, how many sidereal days are in a year on that planet?

One. A sidereal day is the day measured relative to the stars. If you were standing on the midnight side of the planet, it would take 1 full year for you to see the same stars cross your meridian.

CHAPTER 2 The Sky and Telescopes

2.11. If the lunar day were 12 hours long, what would be the approximate time interval between high and low tide?

Since we know that there are two high and two low tides per day, we can divide the length of the day by 4 to get the time between each high and low tide.

$$\text{Time} = \frac{12 \text{ hours}}{4}$$
$$\text{Time} = 3 \text{ hours}$$

2.12. If on a given day, the night is 24 hours long at the North Pole, how long is the night at the South Pole?

0 hours. When the North Pole is completely in darkness, it is winter there. This means that it is summer at the South Pole; so, the South Pole is receiving 24 hours of sunlight. There is no night at the South Pole when the North Pole is completely in darkness.

2.13. On what day of the year are the nights longest at the equator?

This is a trick question. The nights at the equator are **always** 12 hours long.

2.14. How many degrees (°), arc minutes ('), and arcseconds (") does the Moon move across the sky in 1 hour? How long does it take the Moon to move across the sky a distance equal to its own diameter?

The lunar day is 24 hours and 48 minutes long. Therefore the Moon moves 360 degrees (a full circle) in 24.8 hours. Each hour, the Moon moves through an angle, a:

$$a = \frac{360°}{24.8 \text{ hours}}$$
$$a = 14.5°/\text{hour}$$

The Moon moves 14°, 30′, and 0″ every hour.

The Moon's diameter is about 0.5°, or 30′. So to find out how long it takes to travel 0.5°, divide the angular distance traveled by the angular velocity,

$$t = d/v$$
$$t = \frac{0.5°}{14.5°/\text{hour}}$$
$$t = 0.034 \text{ hour}$$
$$(1 \text{ hour} = 60 \text{ minutes})$$
$$t = 2.1 \text{ minutes}$$

So the Moon moves across the sky a distance equal to its own diameter every 2.1 minutes, due to the combined motion of the Earth and the Moon.

2.15. From the fact that the Moon takes 29.5 days to complete a full cycle of phases, show that it rises an average of 48 minutes later each night.

This problem is similar to Problem 2.14, but the motion being considered is somewhat different. On an hour-by-hour scale, the motion of the Earth around its own axis dominates your observations of the motion of the Moon. However, over several days, the motion of the Moon around the Earth adds up to a significant effect.

The Moon moves 360° in 29.5 days, so in 1 day, how far does it move?

CHAPTER 2 The Sky and Telescopes

$$a = \frac{360°}{29.5 \text{ days}}$$
$$a = 12.2°/\text{day}$$

Since the Earth turns 360° in 24 hours, what length of time does 12.2° correspond to in the Earth's sky?

$$b = \frac{24 \text{ hours}}{360°}$$
$$b = 0.067 \text{ hours}/°$$
$$(1 \text{ hour} = 60 \text{ minutes})$$
$$b = 4 \text{ minutes}/°$$

So, by traveling 12.2° through the sky in 1 day, the Moon has delayed its rising by 4×12.2 minutes, or 48.8 minutes.

2.16. What is the altitude of Polaris (Fig. 2-7) as seen from 90, 60, 30, and 0 degrees latitude?

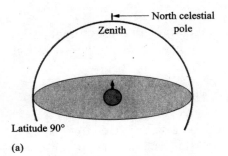
(a)

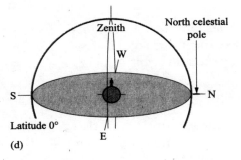
(d)

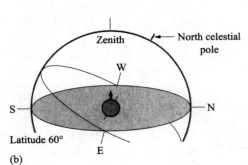
(b)

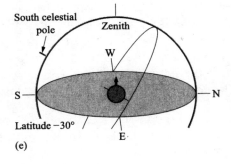
(e)

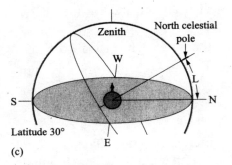
(c)

Fig. 2-7. The altitude of Polaris.

90° latitude is the North Pole. By definition, the North Star is at the zenith, or at 90° altitude.

0° latitude is the equator. At the equator, the North Star is at 0° altitude.

At 60° latitude, the altitude = 60°.

At 30° latitude, the altitude = 30°.

Instrumentation

TELESCOPES

The primary function of a telescope is to gather *more* light than the unaided human eye can gather. The amount of light that can be collected by a telescope is determined by the collecting area (the area of the telescope's lens or mirror that is open to the sky),

$$F = \text{constant} \cdot A$$

Alternatively, the amount of time it takes to collect a certain amount of light scales like the inverse of the area—the bigger the telescope, the less time it takes to collect a certain amount of light,

$$t = \frac{\text{constant}}{A}$$

In theory, this means that more astronomers can use a larger telescope, since it takes less time to accomplish a task. Of course, this is not true in practice. Astronomers observe fainter, more distant objects with larger telescopes, and so take about the same amount of time as on a smaller telescope, where they would restrict themselves to brighter objects.

There are two kinds of optical telescopes: refractors and reflectors. Refracting telescopes use lenses to bend the light to a focus. Reflectors use curved mirrors that reflect the light to the focus (Fig. 2-8).

Reflecting telescopes are preferred over refracting telescopes for several reasons:

1. A large mirror can be as thin as a small mirror, but a large lens must be thicker (thus heavier). For large diameters, lenses get much heavier than mirrors.

2. A lens has two surfaces that must be polished and cleaned; a mirror has only one.

3. Glass absorbs light. The thicker the glass, the more light gets absorbed.

4. Lenses can be supported only around the outside, but mirrors need supporting all across the back.

5. For large lenses, the glass deforms under its own weight and the image slides out of focus.

6. In a lens, different colors are refracted by different amounts. The blue light gets bent further than the red light. This phenomenon is called **chromatic aberration**, and in some cases is a positive quality: prisms can spread out

CHAPTER 2 The Sky and Telescopes

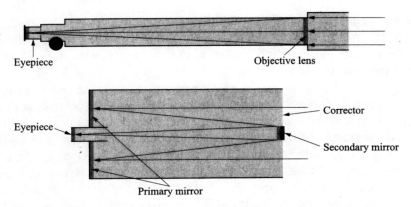

Fig. 2-8. The two types of optical telescopes: refractors and reflectors.

light into a rainbow. However, in the case of a telescope, the light coming through a lens will be focused at different locations, depending on the color. Only one color will be focused, and the others will be out of focus. This can be fixed by combining lenses made out of a number of different materials, which compensate each other. These combination lenses are heavier and prohibitively expensive, especially in larger sizes.

Of course, reflectors are not perfect either. The worst problem is that the outer edges of the image are blurred. This effect is called **coma**. Some of the incoming light rays enter the telescope at an angle, rather than parallel to the axis, and so are imperfectly focused by the mirror. This problem can be fixed with a special lens, but then a special camera with curved photographic plates is necessary to "see" the picture. You can't just look through the telescope with your eye.

The angular resolution of a telescope indicates how close together two points can be before they can't be distinguished. This number should be as small as possible. The angular resolution is measured in arcseconds (an arcsecond is $1/3,600°$). The angular resolution is given by

$$AR = 250{,}000 \cdot \frac{\lambda}{d}$$

where λ is the wavelength at which you are observing and d is the diameter of your telescope (these must be in the same units). If the diameter of the telescope is large, the angular resolution is small. If the wavelength is large, the angular resolution is large. Fortunately, it is easy to build large telescopes for observing at large wavelengths. Unfortunately, at small wavelengths, such as ultraviolet, X-ray, or gamma ray, it is very difficult to build telescopes with even modest diameters. The atmosphere is opaque at these wavelengths, so telescopes must be placed in orbit to be useful. This places strong constraints on the possible size of these telescopes.

Even at relatively long wavelengths, such as in the visible or radio region of the spectrum, orbiting telescopes are preferable to ground-based telescopes. This is because the temperature of the atmosphere is not the same throughout, which causes changes in the density and flow of air. As light travels through air currents or pockets of high- and low-density air, it is refracted, or bent. As these patterns in the atmosphere change, the light coming into your telescope will be observed in a slightly different place on your detector over time, spreading out the image. The phenomenon of the moving light paths is called atmospheric scintillation, and limits the angular resolution of even large telescopes.

Relatively recent developments in telescope technology, such as interferometry or adaptive optics, are able to produce images with angular resolutions comparable to those that could be achieved with space-based telescopes. Interferometry has been used primarily in the radio and millimeter wavelengths, and links many smaller telescopes together to simulate a much larger telescope. Optical interferometers have been built but, because of the higher frequencies, are much more difficult to construct on the same scale as radio and millimeter interferometers. New projects are under way to build optical interferometers in space, which may be able to detect Earth-sized planets around other stars. These projects are still very much in the early planning stages, and rely on unproven technologies, but over the next decade or so the feasibility of such telescopes should be determined. Adaptive optics is used in the optical and infrared to correct for the motion of the atmosphere while the observation is taking place. The correction makes use of a bright star close to the target as a reference. The Gemini North telescope in Mauna Kea, Hawaii (a large telescope with an adaptive optics imaging system) has produced images with less than one-tenth of 1 arcsecond angular resolution. (Recall that 1 arcsecond is the angular diameter of a tennis ball 8 miles away.)

MAGNIFICATION

Telescopes gather more light, and this is the primary reason that they are useful. They can also be used to magnify an image. This is done in practice by changing the eyepiece. However, because the same amount of light is used to make a magnified image as was used to make the original image, the magnified image is fainter. Also, the field of view (the area of sky that can be seen through the telescope) is reduced as the image is magnified. In practice, astronomers rarely magnify an image using their telescope. Instead, they take a picture, using either photographic film or a digital camera, and then use an image-processing program to make the image larger. As a practical rule, the useful magnification of an amateur telescope does not exceed 10 times the diameter of the objective in centimeters. For example, a 4-inch reflector (diameter ~ 10 cm) will magnify $10 \times 10 = 100$ at most. At larger magnifications, the image deteriorates.

DETECTORS

The human eye is unparalleled in its range of color sensitivity, sensitivity to dim light, and adaptability. We can't manufacture anything that comes close to the human eye in overall usefulness. When used in connection with the brain, the eye is far and away the most sophisticated imaging system around. Observation of faint objects, which requires collection of light over extended periods of time, as well as image storage, can be achieved by various devices:

CHAPTER 2 The Sky and Telescopes

1. **Analog camera.** A 35-mm camera on the end of a telescope can take great pictures of the night sky. The shutter can be left open for a long time to image faint objects. The disadvantage of an analog camera is that it is awkward to get the pictures into a computer for analysis. Analog cameras have the advantage of being able to take "true-color" pictures.
2. **Digital cameras.** In astronomy, these are usually called CCDs (short for charge-coupled devices). These cameras provide the best link between a computer and the light coming from the sky. The information is digitized while it is taken, so that developing film or scanning photographs is not necessary. In general, digital cameras do not record in color, so that special filters are required to record only one color at a time. These filtered images can be recombined to form color images.
3. **Photometer.** A photometer adds up the light in an area of an image. All of the spatial information is lost. This is useful when observing objects which change in brightness over time, or when looking at sources whose light cannot be turned into an image (gamma-ray sources, for example).
4. **Spectrometer.** A spectrometer works like a prism, and records a spectrum of an object, in a particular set of wavelengths, with a particular frequency resolution. You will see why this detector is so important if you recall from Chapter 1 all the things we can find out from a spectrum.

Solved Problems

2.17. What size eye would be required to see in the radio bands with the same angular resolution as you have with your present eyes (AR $= 1/60° = 1' = 60''$)?

A typical radio wavelength (from Fig. 1-3) is 10^{-1} m. We can use the angular resolution equation:

$$\text{AR} = 250{,}000 \cdot \frac{\lambda}{d}$$
$$d = 250{,}000 \cdot \frac{\lambda}{\text{AR}}$$
$$d = 250{,}000 \cdot \frac{0.1\,\text{m}}{60''}$$
$$d = 416\,\text{m}$$

This is just the pupil: you would need two of these to have adequate depth perception, and your head would have to grow proportionately to support these eyes. This should explain one reason that we don't see in the radio, even though the atmosphere is transparent to radio waves.

2.18. What is the difference between a photometer and a camera?

A photometer collects all the light as though it came from the same place. There is no spatial information included. A camera, however, includes information about the distribution of the light emitted by the object being observed.

2.19. What is the difference between a spectrometer and a camera?

A spectrometer spreads out the light according to wavelength or frequency, producing a spectrum. A camera spreads out the light according to its position on the sky, producing an image.

2.20. An 0.76-meter telescope can collect a certain amount of light in 1 hour. How long would a 4.5-meter telescope need to collect the same amount of light?

The time required for a telescope to collect a given amount of light is inversely proportional to the area, so we can set up a ratio (as in Problem 1.11) to solve this problem.

$$\frac{T_{4.5}}{T_{0.76}} = \frac{\frac{\text{constant}}{A_{4.5}}}{\frac{\text{constant}}{A_{0.76}}}$$

$$T_{4.5} = \frac{A_{0.76}}{A_{4.5}} \cdot T_{0.76}$$

$$T_{4.5} = \frac{\pi R_{0.76}^2}{\pi R_{4.5}^2} \cdot T_{0.76}$$

$$T_{4.5} = \frac{R_{0.76}^2}{R_{4.5}^2} \cdot T_{0.76}$$

$$T_{4.5} = \frac{(0.76/2)^2}{(4.5/2)^2} \cdot T_{0.76}$$

$$T_{4.5} = \frac{0.144}{5.06} \cdot 1 \text{ hour}$$

$$T_{4.5} = 0.028 \text{ hour}$$

$$(1 \text{ hour} = 60 \text{ minutes})$$

$$T_{4.5} = 1.7 \text{ minutes}$$

The saving in time is substantial. A 4.5-meter telescope can collect as much light in 1.7 minutes as a 0.76-meter telescope can collect in 1 hour.

Supplementary Problems

2.21. How much more light can an 8-meter telescope collect than a 4-meter telescope (in the same amount of time)?

Ans. Four times as much

CHAPTER 2 The Sky and Telescopes

2.22. When the Moon and the Earth become locked in completely synchronous rotation ("tidally locked"), how many orbits will the Moon make in one Earth day?

Ans. One

2.23. When the Moon and the Earth become locked in synchronous rotation, will lunar observers see the Earth pass through phases? will Earth observers see the Moon pass through phases?

Ans. Yes, yes

2.24. If the moon is on the eastern horizon at midnight, what phase is it in?

Ans. Third quarter

2.25. If the rotation axis of the Earth were located in the plane of the orbit, how long would the day be in the summer?

Ans. 24 hours

2.26. Suppose you observe a star directly overhead at 10 p.m. How far to the east or west is it at 11 p.m.?

Ans. 15.04° to the west

2.27. How much longer will the solar day be in the year 3000?

Ans. 0.015 seconds

2.28. Suppose that a solar eclipse occurred last month. Will you observe one this month?

Ans. No

2.29. In what year (approximately) will Vega again become the "North Star"?

Ans. 16,000 C.E.

2.30. What is the advantage of an analog camera over a digital camera?

Ans. Color

2.31. What is the ratio of the flux hitting the Moon during the first quarter phase to the flux hitting the Moon near the full phase?

Ans. 1.005

CHAPTER 3

Terrestrial Planets

The four terrestrial planets are Mercury, Venus, Earth, and Mars. The properties of the surfaces of these planets, and the basic geologic processes that produce them depend on a few basic planetary properties, such as mass, distance from the Sun, and internal composition (see Table 3-1).

Table 3-1. Facts about terrestrial planets

	Mercury	Venus	Earth	Mars
Mass (kg)	0.328×10^{24}	4.87×10^{24}	5.97×10^{24}	0.639×10^{24}
Radius (m)	0.244×10^7	6.052×10^7	6.378×10^7	0.339×10^7
Density (kg/m^3)	5,400	5,200	5,500	3,900
Avg. surface temp. (K)	400	730	280	210
Albedo	0.06	0.65	0.37	0.15
Orbital radius (m)	57.9×10^9	108×10^9	150×10^9	228×10^9
Orbital period (days)	87.97	224.7	365.3	687.0
Orbital inclination (°)	7.00	3.39	0.00	1.85
Orbital eccentricity	0.206	0.007	0.017	0.093
Rotation period (days)	58.65	243.02	1.00	1.03
Tilt of rotation axis (°)	2	177	23.5	25.2

All of the terrestrial planets have undergone significant evolutionary changes over the course of their histories. At one time, all of these planets were as heavily cratered as the Moon, but geologic and atmospheric processes have erased many of these craters. We now understand that even the atmospheres change dramatically over time, as gases are liberated or trapped in rocks or oceans. Studying all of the terrestrial planets is important because it lends insight into the processes governing the Earth's composition, appearance, and evolution.

Formation of Terrestrial Planets

Planetary systems form from the accretion disks around young stars (see Chapter 6 for more on the formation of stars and disks). The gas near the central star remains at a higher temperature than the gas far from the central star. Since different substances condense at different temperatures (see Table 3-2), we would expect to see some substances only in regions far from the central star, such as ices composed of frozen ammonia or methane. In the case of our solar system, this is certainly true. The terrestrial planets in the inner solar system are composed of silicate/iron particles, with very little ice, while solid bodies which orbit the Jovian planets in the outer solar system are composed mainly of ice-shrouded particles.

Table 3-2. Condensation temperatures for different substances

Substance	Condensation temperature (K)
Metals	1,500–2,000
Silicates	1,000
Carbon-rich silicates	400
Ices	200

Once some of these tiny particles have condensed, they begin to **accrete**; i.e., they begin to stick together, and form larger particles. The difference between condensation and accretion is that condensation is a change of state—water vapor condenses on a cold glass in the summer. Accretion is not a change of state, but rather a change in the distribution of particles in space—dust accretes onto furniture. When the particles achieve sizes between a few millimeters and a few kilometers, they are called **planetesimals**. The larger objects accumulate mass quickly, because they have more area, as well as more mass (therefore more gravity).

Calculations of the early solar system show that once planetesimals reach about 1 km in size, they grow through interaction with other planetesimals—sometimes the collisions pulverize both objects, turning them back into many smaller particles, and sometimes the planetesimals combine to form a new, much larger

CHAPTER 3 Terrestrial Planets

planetesimal. These interactions form objects of 1,000 km or more in size. These objects are primarily held together by their own gravity, and are massive enough that gravity regularizes their shapes, making them spherical, rather than potato-shaped like asteroids. Once the planetesimals reach this size, it is difficult to break them up again, and they simply sweep up the rest of the small planetesimals nearby. Many of the intercepted planetesimals form craters on the surface of the forming planet. Some of these types of craters are still visible on the surfaces of the Moon and Mercury. The period of time during which these craters formed is called the "bombardment era," because small objects were constantly impacting the young planets.

When the solar wind began (Chapter 6), it swept out the remaining gas from around the planets, in essence evacuating the area to the current density. This effectively ended planet formation, since there is very little material left to accumulate into planets. The entire process, from accretion disk to planet formation occurred in only a few hundred thousand years.

Solved Problems

3.1. Why are the terrestrial planets spherical in shape?

The terrestrial planets are spherical because of gravity. Gravity pulls towards the center of the object, and once the force is strong enough, it can force even rock to distribute itself spherically.

3.2. Why is there so little ice in the terrestrial planets (compared to the satellites of Jovian planets)?

Different elements have different condensation temperatures. The inner solar system was too hot for ices to condense, and so they did not form here. Ices could not form until the particles were in a much cooler part of the disk, and therefore much further from the Sun.

3.3. Why were impacts so much more common in the past than they are today?

In the early solar system, there was still much debris, and many planetesimals. Since that time, most of these particles have accreted onto the larger planets and moons, so that the probability of intercepting one has become very low.

3.4. How many 100-km diameter planetesimals are needed to form an Earth-size planet? (Assume the planetesimals are spherical, and rocky, so that their density is 3,500 kg/m^3.)

The mass of an object is given by:

$$M = \rho \cdot V$$

where ρ is the density and V is the volume. The volume of a 50-km (spherical) object is

$$V = \frac{4}{3}\pi \cdot R^3$$
$$V = 5.2 \times 10^5 \text{ km}^3$$

Given that the density of the planetesimal is 3,500 kg/m^3, the mass of the planetesimal is

$$M = \rho \cdot V$$
$$M = 3{,}500 \text{ kg/m}^3 \cdot 520{,}000 \text{ km}^3$$
$$M = 2 \times 10^9 \text{ kg/m}^3 \cdot \text{km}^3$$

Convert kilometers to meters:

$$M = 2 \times 10^9 \, \frac{\text{kg}}{\text{m}^3} \cdot \text{km}^3 \cdot \left(\frac{1{,}000 \text{ m}}{\text{km}}\right)^3$$
$$M = 2 \times 10^9 \, \frac{\text{kg}}{\text{m}^3} \cdot \text{km}^3 \cdot 1 \times 10^9 \, \frac{\text{m}^3}{\text{km}^3}$$
$$M = 2 \times 10^{18} \text{ kg}$$

We now have the mass of an individual planetesimal. How many of these do we need to make the mass of the Earth? To find out, divide the mass of the Earth by the mass of a planetesimal.

$$\text{Number} = \frac{M_{\text{earth}}}{M_{\text{planetesimal}}}$$
$$\text{Number} = \frac{6 \times 10^{24} \text{ kg}}{2 \times 10^{18} \text{ kg}}$$
$$\text{Number} = 3 \times 10^6 = 3{,}000{,}000$$

Therefore, it takes roughly 3 million rocky planetesimals of diameter 100 km to equal the mass of the Earth.

3.5. Imagine a large planetesimal, about 100 km in diameter, orbiting the early Sun at a distance of 1 AU. How fast is this planetesimal traveling?

Using Kepler's third law, which states that

$$P(\text{years})^2 = a(\text{AU})^3$$

We find that the period of the planetesimal is 1 year, during which period it travels once around the Sun, or a total distance of $2\pi \cdot r$. Since the radius of the orbit is 1 AU (1.5×10^{11} m), the total distance covered in 1 year is 9.4×10^{11} m. One year is 3.16×10^7 seconds, so the velocity is

$$v = d/t$$
$$v = \frac{9.4 \times 10^{11} \text{ m}}{3.16 \times 10^7 \text{ s}}$$
$$v = 2.98 \times 10^4 \text{ m/s}$$

3.6. Suppose that the gas which formed the planets had cooled much faster, so that the temperature of the gas was below 1,000 K before condensation began. How would the terrestrial planets be different?

CHAPTER 3 Terrestrial Planets

The temperature of the gas affects the condensation of the elements in the gas. If the gas were cooler when condensation began, particles now common in the outer solar system would also be common in the inner solar system. This could include condensates such as ices of methane and ammonia.

Evolution of the Terrestrial Planets

During the bombardment era, the impacts of planetesimals kept the young planets hot. So hot, in fact, that they were completely molten. During this time, the planets **differentiated**. That is, the material separated, with the denser, heavier materials (such as iron) sinking to the center, and the lighter materials (such as rock) floating on top. This led to a layering of materials into three distinct bands: the **core**, mostly iron and nickel; the **mantle**, denser rocks; and the **crust**, lighter rocks. The top of the mantle and the crust form a layer called the **lithosphere**. This layer is formed of relatively rigid rock. Beneath the lithosphere, the rock deforms easily although, strictly speaking, it is not molten. Only on the Earth is this lithosphere broken into plates, which slide on the soft mantle rock below, leading to **plate tectonics**.

The strength and thickness of the lithosphere determine which geological processes operate on the surface. A thick lithosphere suppresses volcanic activity and tectonics. The size and temperature of the interior of the planet governs the thickness of the lithosphere. Hot interiors keep the mantle fluid quite far from the core, so the lithosphere is thin. Planets with cool interiors have thick lithospheres.

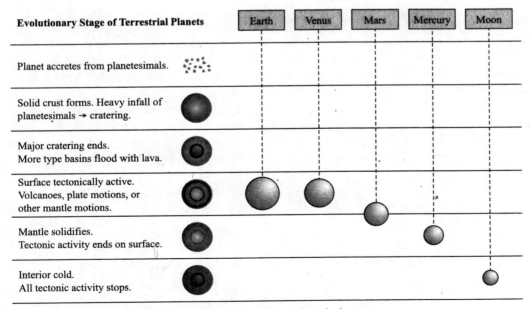

Fig. 3-1. Sequence of terrestrial planet formation and evolution.

All planets are gradually releasing heat into space (see Fig. 3-1). Most of the heat created during the formation of the planets, as planetesimals smashed into each other, has already been lost. Planets currently releasing heat to space are generating it in their cores, as radioactive materials decay. This heat is then either **conducted** or **convected** to the surface, where it is subsequently radiated away. Conduction occurs when molecules exchange energy, so that the interior molecules do not directly carry the energy to the surface, but rather pass it along to other molecules. Convection occurs when the molecules themselves migrate; warm material rises, and cools, and then falls.

PLANETARY SURFACES

Four processes shape and change planetary surfaces: impact cratering, volcanism, tectonics, and erosion. **Impact cratering** occurs on all planets, when objects crash into the planet and leave bowl-shaped depressions (craters) in the surface. **Volcanism** occurs on planets with thin lithospheres, and is the release of molten rock onto the surface of the planet (through volcanoes). **Tectonics** is also more common on planets with thin lithospheres, and is the deformation of the planet by internal stresses. Mountains such as the Appalachian Mountains on Earth, or valleys and cliffs such as the Guinevere Plains on Venus are examples of structures formed by tectonics. (Plate tectonics, the motion of continental plates, is a special case of tectonics, and has only been observed on the Earth so far.) **Erosion** is a gradual alteration of the geological features by water, wind, or ice.

PLANETARY ATMOSPHERES

The terrestrial planets are too small to have captured their atmospheres directly from the accretion disk. These atmospheres were formed during episodes of volcanic activity, which released gases from the molten interior. This is called **outgassing**. Infalling comets or icy planetesimals may have added to these atmospheres, although this contribution is probably small.

The composition of a planet's atmosphere has a dramatic effect on the evolution of that atmosphere. The composition governs the ability to hold heat, and the ability to lose heat. Lighter planets can retain atmospheres made of heavier molecules (see Chapter 1 on escape velocity and the average speed of particles in a gas).

CHAPTER 3 Terrestrial Planets

Solved Problems

3.7. What determines the thickness of the lithosphere of a planet? Why is the thickness of the lithosphere important to the geologic evolution of the planet?

The internal temperature plays a major role in determining the thickness of the lithosphere. Hot planets have thin lithospheres, and cool ones have thick lithospheres. The thickness of the lithosphere determines how geologically active the surface is. If the lithosphere is very thick, volcanoes will never occur, for example.

3.8. The bombardment era completely covered the surfaces of all the solid planets with craters. This era ended about 3.8 billion years ago. Since then, the impact rate has been relatively constant. But some places on the Moon are relatively smooth, with only a few impact craters. Explain how we can determine the age of the "smooth" surface from this information.

Since the surfaces are smooth, they must have been laid down **after** the bombardment era, otherwise they would be as cratered as the rest of the Moon's surface. By counting the number of craters per square meter, and dividing by the flux of impactors (the number per square meter per time), we find the time it has taken to make that many craters on the surface. This must be the age of the "smooth" surface.

3.9. Explain why Mars has no currently active volcanoes.

Mars is much smaller than the Earth, and therefore cooled much more rapidly, decreasing both the temperature and the pressure in the core, which remained at the same volume. Mars no longer has a molten core, and therefore has no internal heat to keep the internal pressure high and drive volcanic activity.

3.10. Why is the Moon heavily cratered, but not the Earth?

Since both of these objects have been around since before the end of the bombardment era, we would expect that they would have been impacted with an equal flux of impactors, and therefore should have comparable crater densities. However, the Earth has an atmosphere, and large amounts of liquid water on the surface, which have erased the craters through erosion.

3.11. If water did not condense out in the inner solar system, where did Earth's oceans come from?

There are two possibilities: first, the water may be the result of small amounts of water that were trapped in the condensing rock as the system formed; the second possibility is that the water has been carried in from the outer solar system by other bodies, comets perhaps, that impacted the Earth.

Mercury

Mercury is the smallest of the terrestrial planets, more comparable in size to the Moon than to the Earth. It is also the closest planet to the Sun, with an orbital radius of only 0.39 AU. Mercury is so close to the Sun that ordinary Newtonian gravity is not sufficient to describe its orbit. The perihelion of Mercury's orbit advances by 574 arcseconds each century. This precession of Mercury's orbit was one of the first successful tests of Einstein's Theory of General Relativity (a more complete theory of gravity).

The rotation period (sidereal day) of Mercury is 58.65 days, exactly 2/3 of the orbital period (87.97 days). When the period and the orbit can be related to each other by a simple ratio of integers in this way, it is called a **resonance**, and is extremely stable. Resonances are the result of gravitational interactions. In this case, Mercury bulges to one side, and when the planet is closest to the Sun, the bulge tries to align itself with the Sun. Over time, internal friction has slowed the rotation of Mercury, so that each time it reaches perihelion, the bulge axis points to the Sun.

The same tidal forces that altered the rotation are also making the orbit less elliptical. Eventually, Mercury will be tidally locked to the Sun, so that one side is always in sunlight, and the opposite side is always in darkness.

Mercury is heavily cratered, like the Moon, but also contains large patches of craterless terrain. This implies that the surface of Mercury is younger than the surface of the Moon. Perhaps Mercury remained geologically active longer because it is larger than the Moon and closer to the Sun. Cooling of the interior has caused the crust to crack and shift vertically, producing "scarps," which are cliffs several kilometers high and hundreds of kilometers long. A particularly spectacular impact left the Caloris Basin, a large impact crater about 1,300 kilometers in diameter, surrounded by circular ripples. On the opposite side of Mercury from the Caloris Basin, the surface topography is dramatically disturbed, apparently by the shock waves of the impact. This region is called the weird terrain.

The density of Mercury is greater than the density of the Moon ($\rho_{Mercury} = 5,420 \text{ kg/m}^3$, $\rho_{Moon} = 3,340 \text{ kg/m}^3$), so it must have more heavy elements than the Moon. One explanation is that during a collision with a large planetesimal, part of the mantle was blown away.

Very little of Mercury's primordial atmosphere remains. The current atmosphere is very thin (less than 10^6 particles/cm^3), and mainly formed from solar wind particles trapped by Mercury's magnetic field, and from recently released atoms from the surface. Where did the magnetic field come from? Mercury's core is solid, so it cannot have a magnetic field for the same reason that the Earth or the Sun do. Probably, the magnetic field is a remnant, frozen in the rock from an earlier time when Mercury's core was molten. This magnetic field is about 100 times weaker than the Earth's.

While the equatorial temperature on Mercury is very high, about 825 K, the temperature at the poles is much cooler, about 167 K (60 K in the shade). Since the atmosphere is very thin, heat does not transfer easily from the equator to the

CHAPTER 3 Terrestrial Planets

poles, and so the temperature differential persists, allowing polar caps, possibly formed of water ice, to survive.

Solved Problems

3.12. Why does Mercury have no significant atmosphere?

Mercury is both small (low mass), and close to the Sun, so the thermal speed of the particles in the atmosphere is high. Recall the escape velocity equation from Chapter 1,

$$v_e = \sqrt{\frac{2GM}{d}}$$

Since the mass of the planet, M, is low, the escape velocity is low, and it is relatively easy for molecules to be going fast enough to escape the gravity of Mercury.

3.13. Your friend claims to see Mercury in the sky at midnight. How do you know he's wrong?

In order to see an object in the sky at midnight, it must make an angle of at least 90° with the Sun (see Fig. 3-2). Mercury is very close to the Sun, 0.39 AU. Even when it is farthest from the Sun, the angle between Mercury and the Sun is 28°, quite a bit less than 90°. So your friend could **never** see Mercury (or Venus!) in the sky at midnight.

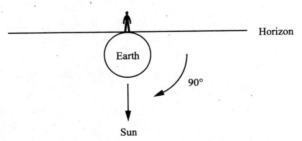

Fig. 3-2. An object in the sky at midnight must make an angle of at least 90° with the Sun.

3.14. How long did astronomers have to wait for the pulse to return? (assume the Earth and Mercury were at closest approach).

At closest approach, the distance between Earth and Mercury is $1 - 0.39 = 0.61$ AU. Converting this to kilometers (multiply by 1.5×10^{11} km/AU) gives 9.15×10^{10} km. The pulse travels at the speed of light, so

$$t = \frac{d}{v}$$

$$t = \frac{9.15 \times 10^{10}\,\text{m}}{3 \times 10^8\,\text{m/s}}$$

$$t = 305\,\text{s}$$

But there are 60 seconds in a minute, so

$$t = 5.1\,\text{minutes}$$

The time for the pulse to travel to Mercury is 5.1 minutes. But the astronomers also had to wait for the pulse to come back, so they had to wait a total of 10.2 minutes.

3.15. The rotation period of Mercury was first determined by bouncing a radar pulse off the surface, and measuring the Doppler shift. Draw a diagram showing how this works (Fig. 3-3).

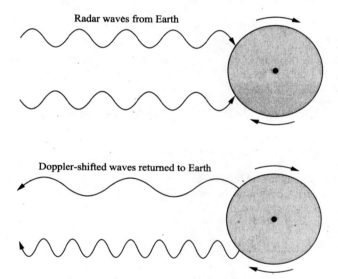

Fig. 3-3. Radar waves are Doppler shifted when they bounce off Mercury.

3.16. Why do astronomers think that Mercury must have a metallic core?

Mercury's density is high compared with the Moon: it is more comparable to the density of the Earth. Therefore, it cannot be rock all the way to the center. In addition, there is a magnetic field around Mercury, which is probably a remnant of a time when it had a molten metallic core. This core might no longer be molten, but it is still metallic.

3.17. How many of Mercury's sidereal days are there in a Mercury year?

Since the rotation period of Mercury is two-thirds the orbital period, the day is two-thirds of the year. There are 1.5 Mercury days in the Mercury year. It helps to draw a diagram (Fig. 3-4).

CHAPTER 3   Terrestrial Planets

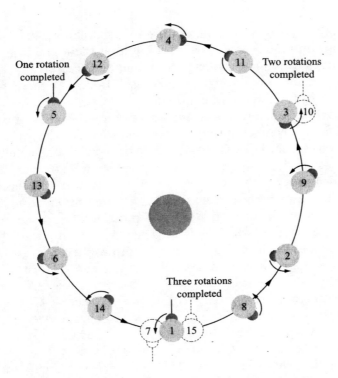

Fig. 3-4. Diagram of the number of Mercury days in a Mercury year.

Venus

In terms of mass and radius, Venus is the most similar to Earth. Venus is about 0.82 Earth masses, and 95% the radius of the Earth. Here, however, the resemblance stops. Perhaps the most poorly understood difference between Venus and the Earth is the sense and rate of rotation. Venus's rotation is retrograde (opposite, i.e., clockwise rather than counterclockwise) to its orbit, and to the orbits and rotations of the vast majority of the rest of the solar system bodies. The rotation is also very slow; Venus's day is about 243 Earth days. One possible explanation is that the young Venus was struck by a large planetesimal, which turned it over, and slowed its rotation. There is no direct evidence to support or refute this hypothesis.

The atmosphere rotates in the same direction as the surface of the planet. Near the surface of Venus, there is very little wind. However, the upper atmosphere is a super-rotator. The upper atmosphere reaches speeds of 100 m/s near the equator, orbiting the planet in only 4 Earth days. On Earth, thin streams of atmosphere (the jet streams) reach comparable speeds, but the bulk of the atmosphere moves much more slowly.

The atmosphere on Venus consists of carbon dioxide (96%) and nitrogen (3%). The atmosphere is so thick that the pressure at the surface of Venus is 90 times the atmospheric pressure on the surface of the Earth. This is roughly equivalent to the pressure 0.5 mile under the surface of an ocean on Earth. The high pressure (plus the CO_2 atmosphere) drives the temperature to a scorching 740 K, which remains quite uniform around the planet. Lead is a liquid at these pressures and temperatures, which makes it quite a challenge to design a spacecraft that can land there and remain functional long enough to take useful data. Higher up in the atmosphere, clouds of sulfuric acid are common.

CO_2 is a greenhouse gas. It reflects infrared radiation in the atmosphere toward the surface, where it becomes trapped. This keeps heat in the atmosphere, rather than releasing it to space. The CO_2 atmosphere on Venus is the result of a runaway greenhouse effect. Early in its history, Venus probably had about the same proportion of water and CO_2 that the Earth did at that time. But because Venus is closer to the Sun, and therefore warmer, the water never condensed out into oceans. The oceans on Earth regulate the amount of CO_2 in the air by trapping carbon deep in the ocean, which gets recycled back into rocks. Without these oceans, the CO_2 on Venus remained in the atmosphere. As volcanoes added CO_2 to the atmosphere, there was nowhere for it to go—it just remained in the atmosphere, further increasing the temperature. As the CO_2 concentration increased, the temperature increased. With no way to remove CO_2 from the atmosphere, the temperature continued to increase as volcanic activity progressed, until the atmosphere reached the hot, dense state it is in now.

Volcanic activity seems to still be a major force on Venus. The entire surface has been repaved by lava flows in the last 500 million years. There are nearly 1,000 volcanoes on the surface of Venus, many of which may still be active, but dormant. No currently active volcanoes have yet been detected.

Solved Problems

3.18. Why is Venus sometimes called the morning (or evening) star?

Venus is never far from the Sun in the sky, and so when it is visible, it appears in the morning or evening, quite close to the Sun. Also, it is one of the brightest objects in the sky, because it is so close to both the Earth and the Sun and has a high albedo (about 0.7). Even when it is farthest from the Sun, the angle between Venus and the Sun is less than 45°.

3.19. Why is Venus's atmosphere so different than Earth's? Explain why this might be considered a "warning" by some scientists.

Originally, these atmospheres were probably similar. But, because Venus was too warm for large liquid oceans, there was no way to scrub CO_2 from the atmosphere. When volcanoes began adding CO_2 to the atmosphere, the lack of a mechanism for recycling CO_2 on Venus made the atmospheric compositions vastly different.

When the atmosphere of Venus began to accumulate high levels of CO_2, infrared radiation (heat) was trapped in the atmosphere, increasing the temperature. Scientists are concerned

CHAPTER 3 Terrestrial Planets

that this could be happening on Earth, with industrial greenhouse gases (such as water vapor and CO_2). If levels of greenhouse gases continue to rise, they may trigger a runaway greenhouse effect of the type that produced Venus's lethal atmosphere.

3.20. How might Venus be different if it were located at 1 AU from the Sun?

If Venus were located 1 AU from the Sun, it would probably be more Earth-like. There would have been large quantities of liquid water on the surface, which would have recycled the CO_2, preventing the runaway greenhouse effect.

3.21. How much more flux (energy per m²) from the Sun does Venus receive every second compared with Earth? Is this difference significant?

Use the inverse square law (Chapter 1). Since we want to compare the flux at Venus and the Earth, we know we want to use a ratio,

$$\frac{F_{Venus}}{F_{Earth}} = \frac{\frac{E_{Sun}}{4\pi d^2_{Sun-Venus}}}{\frac{E_{Sun}}{4\pi d^2_{Sun-Earth}}}$$

$$\frac{F_{Venus}}{F_{Earth}} = \frac{d^2_{Sun-Earth}}{d^2_{Sun-Venus}}$$

$$\frac{F_{Venus}}{F_{Earth}} = \frac{1^2}{0.72^2} = \frac{1}{0.52} = 1.9$$

So the flux at Venus is about twice the flux at the Earth. While this is a significant difference, it is not enough to explain the extreme difference in surface temperature all by itself.

3.22. What is Venus's angular diameter when it is closest? Could you see Venus at this point in its orbit?

Use the small angle formula. The distance between the Earth and Venus when they are closest is $d = (1 - 0.72)\,\text{AU} = 0.28\,\text{AU} = 4.2 \times 10^{10}$ m. From Appendix 2, the diameter of Venus is 12×10^6 m.

$$\theta = 206{,}265 \cdot \frac{D}{d}$$

$$\theta = 206{,}265 \cdot \frac{12 \times 10^6}{4.2 \times 10^{10}}$$

$$\theta = 58.9 \text{ arcseconds}$$

This is very near the limit of the eye's resolution (about 1 arcminute). But that's completely irrelevant, because when Venus is closest, it is in its "new" phase, and just like the new Moon, it is unobservable unless it is actually in front of the Sun (and then you need a special filter to be able to observe it and not be blinded by the Sun).

3.23. How do we know that Venus has been recently resurfaced?

There are few craters on the surface of Venus. Therefore, the surface must have been formed since the age of bombardment. Closer comparisons of the number of craters on Venus to the number of craters in the maria on the Moon, for example, give a better estimate of the age of the surface.

3.24. Why is Venus's rotation considered peculiar?

Venus counter-rotates, or goes the "wrong way." This is peculiar, because the majority of the rest of the planets and their moons all rotate in the same sense. Assuming that Venus originated from the same disk as the rest of the solar system, conservation of angular momentum suggests that the "retrograde" spin of Venus must be the result of a later circumstance that is unique to Venus.

Earth

The Earth is the largest of the terrestrial planets. It is the only one with liquid water on its surface, and the only one that is certain to support life.

The Earth's core is actually two parts: an inner solid core and an outer molten core. The inner core is composed of iron and nickel, while the outer core also contains sulfur. These two cores do not rotate at the same speed. The inner core rotates slightly faster than the outer. As the Earth gradually cools, the inner core grows (Fig. 3-5).

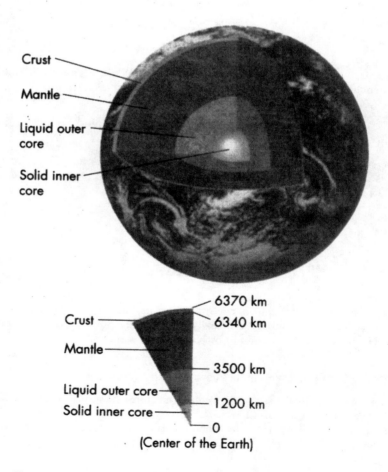

Fig. 3-5. The structure of the Earth.

CHAPTER 3 Terrestrial Planets

The temperature of the core of the Earth is about 6,500 K (for comparison, the Sun's surface is "only" 5,800 K). This core is kept hot by radioactive decay. Convection carries this heat from the deep core through the liquid core, and on through the mantle. These convective motions are very slow in the dense interior of the Earth, but have distinctly observable results: earthquakes, volcanoes, and plate tectonics all result from convection in the Earth's interior.

The rotating, metallic core also creates a magnetic field around the Earth. The overall shape of the field is something like that produced by a bar magnet (Fig. 3-6). The magnetic field traps charged particles from the solar wind. The regions where these particles are trapped are called the **Van Allen belts** (named after their discoverer). Sometimes these particles escape the Van Allen belts and enter the Earth's atmosphere near the North or South Pole. The entry of these high-velocity particles into the atmosphere causes the **aurorae**. The lower regions of the Van Allen belts are about 4,000 km above the Earth, and are primarily populated by protons. The upper regions are populated by electrons, and are located about 16,000 km above the Earth.

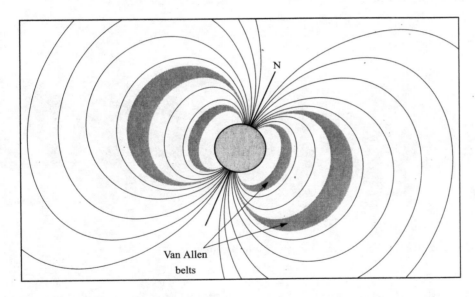

Fig. 3-6. Magnetic field of the Earth.

The magnetic field of the Earth changes with time. It weakens, reverses its North–South polarity, and strengthens again on time scales of about 10^5 years. Currently, the magnetic field is weakening.

The interior of the Earth has been probed by studying earthquakes. Seismologists observe waves traveling through the Earth from the epicenter of the earthquake to other locations. This is much like a medical ultrasound, or sonogram. There are two kinds of waves that travel through the Earth. S waves are transverse: they move material from side to side, perpendicular to the direction they are traveling. P waves are compressional, or longitudinal, like sound: they move material back and forth, in the same direction that they are traveling (Fig. 3-7). S waves do not travel through the bulk of a liquid. They travel only on the surface of a liquid. Ocean waves are a good example of S waves. If the core of the Earth is liquid, we will observe no S waves from an earthquake when we observe from the far side of the Earth. Indeed, this is what is observed, so the core of the Earth is liquid. More detailed analysis has enabled us to map the entire density structure of the Earth, yielding clues about its composition. In particular, the quantity of heavy and radioactive elements can be determined from these data.

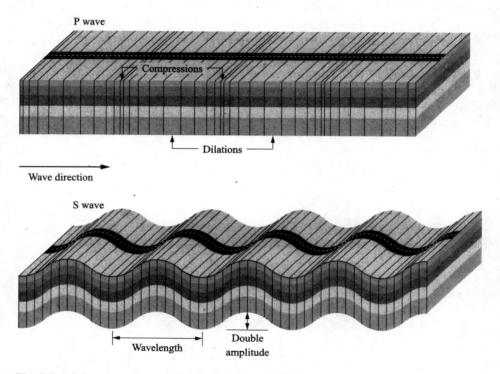

Fig. 3-7. S (transverse) versus P (compressional) waves.

The Earth's crust is made of a series of plates that float on the mantle. As convection cells rise through the mantle, they push against the plates, moving them across the mantle. When plates split apart, or **diverge**, they leave a gap (or **rift**), such as the midocean trenches in the Atlantic. When two plates approach one another, or collide, one plate slides beneath another into the mantle (this is called **subduction**). New material is added to the crust at the midocean ridges, and at volcanic sites, where the mantle wells up to the surface. Sometimes, as plates try to slide past one another, friction holds them still. Pressure builds, until suddenly the plates slip, and an **earthquake** occurs. Geologists are able to measure plate tectonic motion using the global positioning system (GPS), and astronomers are able to measure it using arrays of radio telescopes. The motion is very slow. The Atlantic Ocean, for example, is getting wider at the rate of about 2–3 cm/year.

CHAPTER 3 Terrestrial Planets

Here on Earth, outgassing releases hydrogen, water vapor, CO_2, and nitrogen into the atmosphere. The hydrogen escapes into space, the water vapor condenses and falls into the oceans. The CO_2 is dissolved in the oceans as carbonic acid, which then reacts with silicate rocks to produce carbonate rocks. On the Earth, the atmosphere today is mainly composed of nitrogen (78%), with a significant fraction of oxygen (21%) added by photosynthetic processes in plants. The remaining 1% is mainly argon, an inert (noble) gas. Water, CO_2, and various trace gases and pollutants add up to approximately 0.07% of the atmosphere. Some oxygen molecules are torn apart by ultraviolet (UV) radiation high in the atmosphere. The oxygen atoms recombine with remaining oxygen molecules to form ozone (O_3), which prevents UV solar radiation from reaching the surface. The ozone layer is quite sensitive to destruction by interaction with chlorofluorocarbons (CFCs)—molecules used by humans in refrigeration applications and in propellants, such as in aerosol hairspray or "air fresheners."

The **greenhouse effect** is caused by absorptive molecules in the atmosphere, which keep infrared radiation from escaping the Earth. Most light that falls on the Earth's surface is absorbed, and re-emitted by the Earth as infrared radiation (heat). Molecules such as CO_2 and water trap this infrared radiation. If there are many of these molecules, they can cause a "runaway" greenhouse effect, in which the planet is heated faster than heat can be lost to space. In the case of the Earth, this process would cause the surface to heat up and the water to evaporate, adding more greenhouse gases to the atmosphere, in turn trapping more infrared radiation and evaporating more water, and so on. The feedback loop is predicted to be disastrous. But this scenario considers only one small part of the entire atmosphere–planet interaction. Scientists remain uncertain about other feedback loops. For example, a warmer climate means more trees, which remove CO_2 from the atmosphere, cooling the planet, and perhaps stabilizing the situation. The situation is extremely complicated, and few scientists claim to understand it fully.

Solved Problems

3.25. Compare the Earth's average density with the density of water ($1,000\,kg/m^3$) and rock ($3,500\,kg/m^3$). What can you say about the density of the mantle and the core?

The Earth's average density is about $5,500\,kg/m^3$. This is much higher than the density of rock, the primary substance in the crust, and water, the primary substance on the Earth's surface (the hydrosphere). Therefore, the densities of the mantle and the core must be higher than $3,500\,kg/m^3$ in order for the densities to average $5,500\,kg/m^3$.

3.26. What fraction of the volume of the Earth is contained in the core? What fraction of the mass of the Earth is contained in the core?

At first glance, these seem like the same question, but they are not, because the density of the core is different than the density of the rest of the planet. The fraction by volume is simply calculated by taking the ratio of the volume of the core to the volume of the whole planet:

$$f_{volume} = \frac{\frac{4}{3}\pi \cdot R_{core}^3}{\frac{4}{3}\pi \cdot R_{Earth}^3}$$

$$f_{volume} = \frac{R_{core}^3}{R_{Earth}^3}$$

Using $R_{core} = 1{,}200$ km and $R_{Earth} = 6{,}378$ km,

$$f_{volume} = \frac{(1{,}200 \text{ km})^3}{(6{,}378 \text{ km})^3}$$

$$f_{volume} = 0.007$$

The core of the Earth is 0.007 or about 0.7% of the entire planet, by volume. To find the fraction by mass, we multiply the volume by the density, and then take the ratio,

$$f_{mass} = \frac{\frac{4}{3}\pi \cdot R_{core}^3 \cdot \rho_{core}}{\frac{4}{3}\pi \cdot R_{Earth}^3 \cdot \rho_{Earth}}$$

$$f_{mass} = f_{vol} \cdot \frac{\rho_{core}}{\rho_{Earth}}$$

Using $\rho_{core} = 12{,}000 \text{ kg/m}^3$ and the average density of the Earth, $\rho_{Earth} = 5{,}500 \text{ kg/m}^3$,

$$f_{mass} = 0.007 \cdot \frac{12{,}000 \text{ kg/m}^3}{5{,}500 \text{ kg/m}^3}$$

$$f_{mass} = 0.015$$

The core is 0.015 or about 1.5% of the Earth by mass. This is about twice the fraction by volume.

3.27. What is the difference between ozone depletion and the greenhouse effect?

Ozone depletion is caused by chlorofluorocarbons (CFCs), and allows more ultraviolet light into the atmosphere. The greenhouse effect is caused by greenhouse gases such as CO_2 and water vapor, and keeps infrared light from escaping the Earth's atmosphere. While the effect of depleting the ozone layer might contribute to the infrared radiation coming from the ground (because the ultraviolet gets absorbed by the ground and re-emitted in the infrared), they are not the same physical process, and have quite different origins.

3.28. The Atlantic Ocean is approximately 6,000 km across. How long ago were North America and Europe located next to each other on the planet? (Assume the continents have been drifting apart at the same speed the entire time.)

The Atlantic Ocean is growing by about 2.5 cm/year. To calculate the length of time it took to grow to its current size, divide the width of the ocean by its rate of growth,

$$t = \frac{d}{v}$$

$$t = \frac{6{,}000{,}000 \text{ m}}{0.025 \text{ m/year}}$$

$$t = 240{,}000{,}000 \text{ years}$$

CHAPTER 3 Terrestrial Planets

North America and Europe were located next to each other 240 million years ago. Since that time, they have been gradually drifting apart.

3.29. Compare the ages of surface rocks on the Earth with the accepted age of the Earth (4.5 billion years). The oldest surface rocks are about 3.8 billion years old, and 90% of the surface rocks are less than 600 million years old. How can you reconcile this information?

The age of surface rocks seems to disagree with the accepted age of the Earth, but it is important to remember that the surface of the Earth has been recycled several times since the Earth was formed. The rocks on the surface are "new" rocks. They have been formed only recently by volcanoes or other tectonic processes, and so can not be used to find the age of the Earth, just the age of the surface, which varies with location.

Moon

The Moon is primarily composed of basaltic rock. Because it lacks an atmosphere, or water on the surface, erosion (except erosion by impact of micrometeorites) is an insignificant process on the Moon. The entire history of the surface is preserved.

The interior of the Moon is solid. It has no molten core, and therefore is geologically dead. Quite sensitive seismology equipment, carried to the Moon by Apollo astronauts, detected vibrations caused by tidal interaction with the Earth, and vibrations caused by impact from meteors, but no significant moonquakes, which would indicate a geologically active core and allow the interior to be probed.

The core was active in the past, and lava flowed on the surface when giant impacts cracked the crust. This flowing lava filled in the lowlands and many craters, and created dark, smooth pools of rock, called **maria** (Fig. 3-8). We know that this occurred after the bulk of the craters were formed, because there are far fewer craters in the maria than elsewhere on the Moon.

The highlands, also known as terrae, have a lighter appearance, are richer in calcium and aluminum, and are more heavily cratered (hence older) than the maria. The maria are richer in iron and magnesium and appear darker.

The other common volcanic features on the Moon are **rilles** or **rima**: long thin structures resembling dry riverbeds. One explanation for these features is that lava tubes left by prior activity have collapsed, leaving these long, thin indentations (Fig. 3-9).

Other than the maria, the most conspicuous features on the Moon are craters. There are dozens of craters on the near side of the Moon that can be distinguished even with binoculars. Bright rays often surround these craters. The rays are formed of ejected material from the crater, that is lighter in color than the surrounding region.

Astronomers still debate the origin of the Moon. The current best explanation is that the Earth was impacted by a Mars-sized object about 4.6 billion years ago, and was partially pulverized. A large amount of ejecta was produced, which then

CHAPTER 3 Terrestrial Planets

Fig. 3-8. Image of the Moon. (Courtesy of NASA.)

Fig. 3-9. Plato Rille (or Rima). (Courtesy of NASA.)

CHAPTER 3 Terrestrial Planets

re-accreted to form the Moon. Other theories include simultaneous formation, in which the Moon and the Earth formed at the same time out of the same portion of the disk; gravitational capture, in which the Moon was just passing by, and was captured by the Earth's gravitational field; and a fission theory in which the Moon split off from the Earth.

The Moon rotates on its own axis with a period equal to the orbital period around the Earth. This means that the same side of the Moon faces the Earth at all times. This is called **synchronous rotation**, and explains why the features of the Moon always look the same from Earth. The near side of the Moon has a thinner crust than the far side. Early in the Moon's history, impacting bodies penetrated to the molten mantle on the near side, allowing lava to pool on the surface, and form the maria. The cooled lava is heavier than the crust material. As a result of the Earth's gravitational attraction, the Moon has shifted its mass distribution so that it is actually heavier on the near side; therefore, that side is always falling towards the Earth. A secondary result is that the crust is thicker on the far side of the Moon.

The surface of the Moon is covered with a thick layer of a fine, powdery dust called **regolith**. This dust layer was caused by impacts from micrometeorites, and averages many meters in depth. It is far deeper in the floors of valleys than on the slopes or ridges surrounding them. The famous photograph in Fig. 3-10 shows an astronaut's footprint imbedded in regolith.

The average density of the Moon is about $3,300 \, kg/m^3$, implying that it is primarily composed of rock and does not have an iron core of significant size. Otherwise, the interior of the Moon has not been deeply probed.

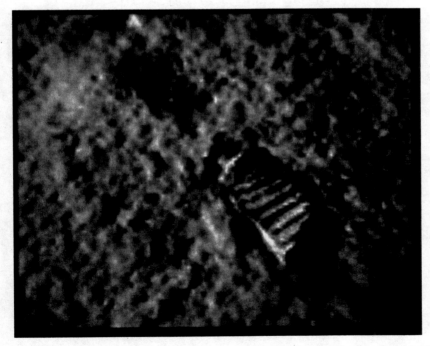

Fig. 3-10. An astronaut's footprint in the lunar regolith. (Courtesy of NASA.)

From radioactive dating of Moon rocks brought back to Earth by astronauts, we have been able to determine the age of the Moon's surface. The maria are younger than the highlands; rocks from the maria formed about 3.8 billion years ago, and highland rocks formed 4.3 billion years ago.

Solved Problems

3.30. Explain the importance of determining the ages of Moon rocks.

The Moon is the closest "pristine" environment in the solar system. Rocks have not been cycled and recycled on the Moon (except in the maria), so the surface rocks have remained unchanged since the Moon was formed, and are the same age as the Moon. Rocks in the maria are more recent, and knowing the age of these rocks gives the age of the maria themselves. Bringing back rocks from the Moon allowed us to date them using radioactive tracers (this is similar to carbon-dating, but uses different elements), and find out how old these surfaces are. In turn, we can compare these regions on the Moon with similar regions on other planets and their moons to figure out how long ago features on other planets formed.

3.31. Does the Moon rotate on its own axis (relative to the stars)?

Yes. This is best explained with a diagram (Fig. 3-11).

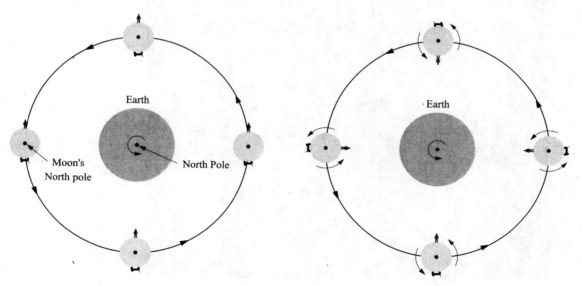

Fig. 3-11. The moon rotates on its axis.

CHAPTER 3 Terrestrial Planets

3.32. How do we know the Moon does not have a large metallic core?

The primary evidence is that the density of the Moon ($\rho_{Moon} = 3{,}340\,\mathrm{kg/m^3}$) is very close to the density of rock. Therefore, the Moon is likely to be mostly made of rock.

The second piece of evidence is that the Moon has no magnetic field, as usually results from a molten iron core. We know the core was molten at some point (because of the maria), but it probably was not made of iron; otherwise, the Moon would probably have a magnetic field like that of Mercury.

3.33. What is the difference between rilles and rays?

Rilles are the result of tectonic activity—probably collapsed lava tubes. Rays, on the other hand, are the result of impacts from incoming bodies. While rilles often meander across the lunar landscape, rays are always relatively straight and always radiate away from a crater.

3.34. What is the difference between maria and the highlands?

Maria are cooled basaltic lava, and are much denser (therefore heavier) than the rock which forms the highlands. Maria have fewer craters and so are presumed younger than the lunar highlands.

3.35. Why has the Moon's interior cooled nearly completely, while the Earth's interior remains hot?

The Moon is much smaller than the Earth, and so has more surface area relative to its volume. This meant more area to radiate away heat, so the Moon cooled faster than the Earth. In addition, the Moon has much less mass, and is of lower density. Therefore, the Moon had less radioactive material to keep the core hot through radioactive decay.

Mars

Mars is the most distant terrestrial planet from the Sun. The mass of Mars is about one-tenth the mass of the Earth, and Mars is about half as large in diameter. The orbital period of Mars is 1.88 years, and so the seasons on Mars are nearly twice as long as the seasons on Earth. These seasons are qualitatively similar to Earth's, since the tilt of Mars's rotation axis is quite similar (25°, compared with the Earth's 23.5°). Models of the evolution of Mars's orbit and tilt show that in the past the tilt has varied erratically between 11° and 49°, which may have had dramatic effects on prior climatic conditions. Like the Earth, Mars has polar caps; unlike the Earth, one of these is mostly CO_2 ice. Temperatures on Mars range from 133 K to 293 K ($-220°F$ to $70°F$), and the atmospheric pressure at the surface is very low, about 1/100 the pressure at the surface of the Earth.

The atmosphere on Mars is 95% CO_2, similar to the atmosphere of Venus. Because the atmosphere is so much thinner, however, there has been no runaway greenhouse effect on Mars. The atmosphere is thick enough to support tenuous water ice clouds, and enormous dust storms, some of which are over a mile high.

Although Mars has no plate tectonic activity, the surface of Mars has been shaped by geologic activity. The Tharsis Plateau contains a string of volcanoes, the largest of which is Olympus Mons. This volcano is over 27,000 meters high (about three times as high as Mt. Everest), with a base 600 km in diameter (an area the size of Utah). Olympus Mons grew to such enormous size because the hole in the crust that brought magma to the surface stayed stationary over geologic time scales. Also spectacular is Valles Marineris, a gigantic rift 3,000 km long and up to 600 km wide. In places this rift is 8 km deep. This feature is not the result of plate tectonics, as it would be on the Earth. Instead, Valles Marineris is the result of the cooling and shrinking crust splitting open at the equator, much like the surface of drying clay.

The red color on the surface of Mars is caused by rust—surface iron has been oxidized. Oddly, though the surface is iron-rich, the interior seems to be iron-poor. Mars has a much lower density than the other terrestrial planets (about 3,900 kg/m^3), and so we suspect it has little iron in its core. This idea is supported by the lack of a magnetic field around Mars.

Recently, Mars Global Surveyor found strong evidence for recent water flows on Mars. Recent water on the surface implies there may be water locked just under the surface as permafrost. The presence of water on Mars might suggest that there has been life on Mars in the past, or that Mars may become capable of supporting life in the future.

Moons of Mars

Mars has two moons, Phobos and Deimos. Both of these moons are quite small, and irregularly shaped, and are probably captured asteroids. Phobos' longest axis is about 28 km long, and Deimos' longest axis is about 16 km long. Both moons are in synchronous rotation about Mars (the same side always faces the planet). Phobos has many craters and deep cracks, while Deimos appears smoother. The orbital radii are 9.38×10^3 km and 23.5×10^3 km and the orbital periods are 7.7 hours and 30.2 hours for Phobos and Deimos, respectively. As the rotational period of Mars is 24.6 hours, an observer on Mars would see Phobos rising in the west and setting in the east.

CHAPTER 3 Terrestrial Planets

Solved Problems

3.36. What might be a result of a sudden melting of the CO_2 ice caps on Mars? (Hint: melting these caps would release large amounts of CO_2 into the atmosphere.)

The sudden addition of a large amount of CO_2 to the atmosphere of Mars would significantly warm the planet. With no water oceans to cycle the carbon back into rocks, it might seem that there is the potential for a runaway greenhouse effect. However, there is not nearly enough CO_2 to actually cause a runaway greenhouse effect. Mars is much less massive than Venus, and so the atmosphere might not remain on the planet, particularly if it becomes very warm (whether or not the atmosphere would "stick" depends on **how** warm it becomes). At least temporarily, the temperature would rise.

3.37. How might you measure the mass of Mars from the orbits of one of its moons?

From Chapter 1, the period, the masses of the orbiting bodies, and the distance between them are all related:

$$P^2 = \frac{4\pi^2 a^3}{G(m + M)}$$

The distance of the moon from Mars (a) can be measured from pictures of the orbit over time. Converting to a linear distance (instead of an angular one) requires knowing the distance to Mars, but the *period* of Mars is observable, and so we could determine the distance from Mars to the Sun from Kepler's third law. If we make our measurements when Mars is opposite the Earth from the Sun, the Earth–Mars distance is easily determined by subtracting the Earth–Sun distance from the Mars–Sun distance.

From the same images of the Moon over time, we can determine the period of the Moon (P). If we assume that the mass of the Moon is small compared to Mars (a good assumption since the moons are so much smaller), we know all the factors in the equation, and can solve for M, the mass of Mars.

3.38. Why are there no volcanoes the size of Olympus Mons on Earth?

There are three contributing factors. First, the continental plates move across the top of the mantle. This means that the holes through the mantle, which allow the very hot magma to come to the surface, are not always in the same location on the plates. On Earth, we are more likely to observe many smaller volcanoes than one huge volcano like Olympus Mons. Indeed, the Hawaiian islands are just such a chain, which develops as the plate slides over the mantle "hot spot." Secondly, the surface gravity on Mars is significantly lower than on Earth, so that taller structures are more stable (less likely to collapse under their own weight) on Mars than on Earth. Thirdly, erosion is much more effective on the Earth than on the surface of Mars. Volcanoes that develop on Earth are worn down again by the action of wind and water.

3.39. Why do astronomers think Phobos and Deimos are captured asteroids?

First, Phobos and Deimos are very asteroid-like. They are rocky, and potato-shaped. Secondly, Mars is located near the asteroid belt. Thirdly, their orbits are unstable. In only 50 million years, Phobos will no longer orbit Mars. This is an astronomically short period of time, and so it must have recently begun to orbit Mars. All of these pieces of information add up to a picture of captured asteroids.

3.40. With how much kinetic energy (KE) would a 1 kg piece of rock have to be traveling in order to leave the surface of Mars as a meteoroid? Compare this to the amount of energy produced by 1 megaton of TNT (4×10^9 joules).

We can use the equation for escape velocity, plus Mars data from Table 3-1, to find the speed such a rock would need to be traveling:

$$v_e = \sqrt{\frac{2GM}{d}}$$

$$v_e = \sqrt{\frac{2 \cdot 6.67 \times 10^{-11} \text{ m}^3/\text{kg/s}^2 \cdot 6.4 \times 10^{23} \text{ kg}}{3{,}393{,}000}}$$

$$v_e = 5{,}020 \text{ m/s}$$

This is about half of the escape velocity from the Earth (11.2 km/s). The kinetic energy is given by

$$KE = \frac{1}{2}mv^2$$

$$KE = \frac{1}{2}(1 \text{ kg})(5{,}020)^2$$

$$KE = 1.3 \times 10^7 \text{ joules}$$

The rock requires 13 million joules of energy. One megaton of TNT could lift only 33 of these rocks (33 kg) from the surface of Mars. This is about half the mass of a person.

Supplementary Problems

3.41. What is the escape velocity from Mercury?

Ans. 4.2 km/s

3.42. What is the average thermal speed of hydrogen atoms near the poles at the surface of Mercury (assume $T = 167$ K; $m_H = 1.6735 \times 10^{-27}$ kg)?

Ans. 3.3 km/s

3.43. Venus's semi-major axis is 0.72 AU. Use Kepler's third law to find its period.

Ans. 0.61 years

3.44. How many Venusian days are in a Venusian year?

Ans. 0.9

CHAPTER 3 Terrestrial Planets

3.45. A tiny new planet, about the size of the Moon, is discovered between the orbits of Venus and Mercury. **Most likely**, this planet will

  (a) orbit clockwise   **(b) orbit counterclockwise**   (c) can't tell
  (a) have a cratered surface (b) have a smooth surface   (c) can't tell
  (a) have a molten core   **(b) have a solid core**   (c) can't tell
  (a) have an atmosphere   **(b) have no atmosphere**   (c) can't tell
  (a) have a magnetic field   (b) have no magnetic field   **(c) can't tell**
  (a) rotate clockwise   **(b) rotate counterclockwise**   (c) can't tell

3.46. What is the equatorial rotational speed of Mercury?

  Ans. 3.03 m/s

3.47. What was the fractional change in wavelength ($\Delta\lambda/\lambda$) of the radio waves bounced off Mercury's approaching limb?

  Ans. 1×10^{-8}

3.48. Suppose that space-traveling seismologists observe S waves from a quake originating on the far side of a planet. What does this tell them about the core?

  Ans. It is solid

3.49. Suppose that the Earth were made entirely of rock. How would it be different?

  Ans. Much weaker magnetic field, no tectonics, no volcanism, possibly no carbon in the atmosphere, no aurorae

3.50. Suppose that you can lift 68 kg on Earth. How much could you lift on the Moon?

  Ans. 408 kg

3.51. How much larger (in volume) is the Earth than the Moon? than Mars?

  Ans. 49.4, 6.6

3.52. Calculate the angular size of the Moon from the Earth using the small angle equation.

  Ans. 0.5°

3.53. What is the angular size of Mars in the sky at its closest (assume circular orbits)?

  Ans. 18″

3.54. Suppose a colony is established on Mars. How long would it take for a Martian doctor to send a question to a colleague on Earth and receive a response? (Assume the colleague knows the answer "off the top of his head.")

  Ans. 8.7 minutes

CHAPTER 4

Jovian Planets and Their Satellites

The Jovian planets are Jupiter, Saturn, Uranus, and Neptune. These are all gas giant planets, with much higher masses and lower densities than the terrestrial planets. None of these planets has a solid surface. All of these planets have a large number of moons and also ring systems. The majority of the planetary mass in the solar system (99.5%) is in the Jovian planets, but they are still only 0.2% of the mass of the Sun. The physical and orbital properties of the Jovian planets are listed in Table 4-1.

All of the Jovian planets are flattened (oblate), so that the diameter from pole to pole is less than the diameter at the equator. This is caused by their size and rapid rotation. Table 4-1 shows that the Jovian planets complete a full revolution in less than 1 Earth day, despite their tremendous size. Remarkable weather patterns are characteristic of Jovian atmospheres and the composition creates colorful effects. The Jovian atmospheres also exhibit enormous storms. This vigorous atmospheric activity is caused by the rapid rotation of the planets and is driven by heat released from their interiors (see discussion below).

All of the Jovian planets formed in the outer part of the solar accretion disk (see Chapters 3 and 6 for more about the disk), and they are considerably richer in light elements than the terrestrial planets. The compositions of the Jovian planets are roughly comparable to that of the Sun. For example, the atomic composition of Jupiter is 90% hydrogen, 10% helium, and less than 1% trace elements. The Sun is 86% hydrogen, 14% helium, and less than 1% trace elements. Uranus and Neptune have higher portions of heavy elements than Jupiter or Saturn, but still they are a small fraction of the total mass. The similarity in composition between the Jovian planets and the Sun supports the idea that the entire solar system formed from the same well-mixed cloud. Most of our knowledge of the Jovian planets and their satellites comes from the Voyager 1 and 2 missions, and additional information about Jupiter and its moons was obtained during the Galileo and Cassini missions.

Table 4-1. Facts about Jovian planets

	Jupiter	Saturn	Uranus	Nepture
Mass (kg)	1.9×10^{27}	5.7×10^{26}	0.87×10^{26}	1.0×10^{26}
Diameter (m)	143×10^6	121×10^6	51×10^6	50×10^6
Density (kg/m^3)	1,300	700	1,300	1,600
Albedo	0.52	0.47	0.51	0.41
Average distance from the sun (m)	0.778×10^{12}	1.43×10^{12}	2.87×10^{12}	4.50×10^{12}
Orbital period (years)	11.86	29.46	84.01	164.8
Orbital inclination (°)	1.31	2.49	0.77	1.77
Orbital eccentricity	0.048	0.056	0.046	0.010
Rotation period (days)	0.41	0.44	0.72	0.67
Tilt of rotation axis (°)	3.1	26.7	97.9	28.3

Jupiter

Jupiter, the largest of the Jovian planets, shows a system of bright and dark stripes. **Belts** are low, warm, dark-colored regions that are falling through the atmosphere. **Zones** are high, cool, light-colored regions that are rising through the atmosphere. Jupiter has ammonia, sulfur, phosphorus, and other trace elements in its atmosphere that contribute variously to the banded structures and the multicolored clouds.

The Great Red Spot on Jupiter (Fig. 4-1) is one such storm, which has persisted since Galileo first observed it nearly 400 years ago. This storm changes in size, but on average is about 1 Earth diameter high, and 2 Earth diameters across. The atmospheric activity is driven by energy produced by the release of gravitational potential energy, as material falls towards the center of the planet. Most of the energy was produced long ago, in the initial contraction of the planet material, and is just now finding its way to the surface. In fact, Jupiter emits twice as much energy as it receives from the Sun.

Jupiter has a dense core of magnesium, iron, silicon, and various ices. Jupiter's core is about 15 times the mass of the Earth. The core has been detected by gravity experiments on fly-by missions such as Voyager. Scientists believe that a thick layer of **metallic hydrogen** surrounds Jupiter's core. Metallic hydrogen is hydrogen that is so dense that the electrons are free to move through it freely. This is similar to the way electrons behave in metals: hence the name. The liquid metallic hydrogen and the rapid rotation of the planet generate the strong magnetic field that surrounds Jupiter.

CHAPTER 4 Jovian Planets and Their Satellites — 79

Fig. 4-1. Jupiter's Great Red Spot, with images of Earth superimposed for scale. Composite from NASA images.

Saturn

Saturn is the second largest of the Jovian planets, and has the lowest density of any planet in the solar system. Saturn has a dense core of magnesium, iron, silicon, and various ices. The core is about 13 Earth masses. Saturn also has a layer of metallic hydrogen surrounding its core, producing a large magnetic field (*metallic hydrogen*: see section on Jupiter, above). The composition of Saturn is similar to that of Jupiter, probably with more hydrogen, to account for the lower density.

Saturn's zones and belts are not as pronounced as Jupiter's. The smeared appearance is due to light scattering from ammonia crystals that form the relatively cold upper atmosphere of the planet. Saturn's atmosphere, like Jupiter's, is heated by energy released from the interior of the planet. The source appears to be gravitational potential energy of helium droplets sinking to the interior.

Uranus

At first, Uranus was thought to be a relatively featureless planet. Voyager images showed a quite plain disk. More recently, astronomers have discovered that this was an accident of the observing season. Voyager passed near to Uranus during a period of few storms and little atmospheric activity. Uranus's rotation axis is tilted nearly 90°, so that it is almost in the plane of the ecliptic. Consequently, one pole faces the Sun for half the orbit, then the other. The sunward side of Uranus absorbs a lot of radiation, which should heat the planet and then drive convection toward the opposite pole. This is not observed at the surface, however, and the heat transport must take place in the deeper layers. Perhaps this mechanism is somehow responsible for keeping the surface of Uranus featureless during certain portions of its orbit. Uranus's atmosphere contains relatively high amounts of methane, which absorbs strongly in the red, causing the reflected sunlight from the planet to appear blue.

Studies of the density and flattening of Uranus indicate that the interior might consist of hydrogen, water, and a small (Earth-sized) core of rock and iron.

Neptune

The atmospheric composition of Neptune is similar to that of Uranus. Compared to Uranus, Neptune appears more blue, presumably due to a higher concentration of methane (2–3%) in the atmosphere. Neptune's Great Dark Spot was first observed by Voyager 2 in the Southern Hemisphere in 1989. This storm disappeared by the time the Hubble Space Telescope observed Neptune in 1994, but a new one had formed in the north by 1995. Neptune releases internal energy, driving supersonic winds to speeds of over 2,000 km/hr. It does not appear that the energy released is a remnant from the formation of the planet, because Neptune is much smaller than Jupiter. The exact source of the released energy is unclear.

Neptune's internal structure might be similar to that of Uranus. The interior consists of hydrogen, water, and a small (Earth-sized) core of rock and iron.

Solved Problems

4.1. How does Jupiter generate its internal heat? How does the Earth generate its internal heat? Why are the two different?

Jupiter generated its internal heat by giving up gravitational potential energy as material fell onto the core. The energy currently being emitted from Jupiter is energy produced in the initial rapid contraction. The Earth generates its internal heat from radioactive elements which decay in the core. The Earth is much smaller than Jupiter, and radiated away all of its gravitational potential energy long ago.

CHAPTER 4 Jovian Planets and Their Satellites

4.2. Why are the Jovian planets different colors?

The color of a Jovian planet is a function of the composition. For example, Uranus and Neptune contain methane, and are blue. These compositions differ for each planet, depending on where they condensed out of the planetary disk in the early solar system. Jupiter has quantities of other molecules, such as ammonia, and is red-orange in color.

4.3. Why is Jupiter brighter than Mars in our night sky, even though it is farther away?

Jupiter has a much larger angular size than Mars:

$$\theta = 206{,}265 \cdot \frac{D}{d}$$

For Jupiter, we find that the angular size is $38''$, while for Mars, the angular size is $6''$. In addition, Jupiter's albedo is higher than that of Mars, so that it reflects more of the sunlight towards the Earth.

4.4. What would happen to the temperature of the Earth's surface if the Sun stopped shining? How about Saturn?

If the Sun stopped shining, the temperature of the Earth would drop dramatically. The Earth does not produce significant radiation from the surface. Saturn, however, produces about half of the energy emitted, and most of that is in the infrared, so the temperature would not drop as fast.

4.5. Which of the outer planets have seasons? Why?

Seasons are a result of an axial tilt. Uranus certainly has extreme seasons! Saturn and Neptune, with axial tilts of $27°$ and $30°$, respectively, also have seasons. Jupiter, however, does not, because Jupiter's axis is tilted only $3°$ with respect to its orbital plane.

Moons

All of the Jovian planets have moons. Some of these are so large that if they orbited the Sun, instead of a planet, they would be considered planets themselves. It is not uncommon for astronomers to discover previously unknown moons of these planets even today. Usually, these newly discovered moons are quite small. Jupiter has 28 known moons, Saturn has 30, Uranus has 21, and Neptune has 8. We will not talk individually about each of these 87 moons. Instead, we will point out some general properties, summarize the moons in Table 4-2, and then discuss only the most interesting few.

Most of the moons of the outer planets have quite low densities, less than 2,000 kg/m^3, implying that they have large fractions of ice, probably water ice. Individual moons formed in place, are captured asteroids, or are remnants of other moons, destroyed in collisions. Triton is the only moon of a Jovian planet that does not fit one of these categories.

Many of these moons are geologically active. Moons can be heated by gravitational contraction (like Jupiter) or by radioactive heating (like the Earth). Moons can also be heated by tidal stresses, exerted by their parent planets. As the moon rotates while it is stretched out of shape by the gravitational force of its parent body, friction develops and heats the interior.

The Galilean moons are the four largest of the Jupiter system, and are named after their discoverer, Galileo. The physical and orbital properties of the largest moons of the Jovian planets are listed in Table 4-2 and other information is given in Table 4-3.

JUPITER'S MOONS

Io. Of the four Galilean moons, Io is the closest to Jupiter. Io is geologically active. There are many active volcanoes on Io, and Fig. 4-2 shows the eruption of one of these, Prometheus. The core of Io is kept active because of the tidal pulls of Jupiter and Europa. Interactions with the more distant moons keep Io's orbit from becoming perfectly circular, and so the interior remains molten, all the way out to a very thin crust.

Table 4-2. Physical properties of the major satellites of the outer planets

Moon	Planet	Diameter (km)	Mass (kg)	Average orbital distance (km)	Period (days)
Io	Jupiter	3,360	8.89×10^{22}	4.22×10^{5}	1.77
Europa		3,140	4.85×10^{22}	6.71×10^{5}	3.55
Ganymede		5,260	1.49×10^{23}	1.07×10^{6}	7.16
Callisto		4,800	1.07×10^{23}	1.88×10^{6}	16.69
Rhea	Saturn	1,530	2.45×10^{21}	5.27×10^{5}	4.52
Mimas		380	3.82×10^{19}	1.86×10^{5}	0.94
Dione		1,120	1.03×10^{21}	3.77×10^{5}	2.74
Tethys		1,050	7.35×10^{20}	2.95×10^{5}	1.89
Iapetus		1,440	1.91×10^{21}	3.561×10^{6}	79.3
Enceladus		500	8.09×10^{19}	2.38×10^{5}	1.37
Titan		5,150	1.35×10^{23}	1.222×10^{6}	15.95
Oberon	Uranus	1,520	2.94×10^{21}	5.83×10^{5}	13.46
Titania		1,580	3.45×10^{21}	4.36×10^{5}	8.72
Umbriel		1,170	1.25×10^{21}	2.66×10^{5}	4.15
Ariel		1,160	1.33×10^{21}	1.91×10^{5}	2.52
Miranda		470	7.35×10^{19}	1.30×10^{5}	1.41
Triton	Neptune	2,700	2.13×10^{22}	3.55×10^{5}	5.88

CHAPTER 4 Jovian Planets and Their Satellites

Table 4-3. Major satellites of the outer planets: quick information table

Moon	Planet	Current geologic activity	Past geologic activity	Craters	Tidally locked	Atmosphere	Ice content
Io	Jupiter	Yes	Yes	No	No	Thin SO_2	Low
Europa		Yes?	Yes	Few	No	No	15%
Ganymede		No	Yes	Yes	Yes	No	45%
Callisto		No	No	Yes	Yes	No	50%
Rhea	Saturn	No	No	Yes	Yes	No	> 50%
Mimas		No	No	Yes	Yes	No	> 50%
Dione		No	Yes	Yes	Yes	No	> 50%
Tethys		No	Yes	Yes	Yes	No	> 50%
Iapetus		No	No	Yes	Yes	No	> 50%
Enceladus		?	Yes	Yes	Yes	No	> 50%
Titan		?	?	Yes	Yes	N_2	50%
Oberon	Uranus	No	Yes	Yes	Yes	No	> 50%
Titania		No?	Yes	Fewer	Yes	No	> 50%
Umbriel		No	Yes	Yes	Yes	No	> 50%
Ariel		No?	Yes	Fewer	Yes	No	> 50%
Miranda		No?	Yes	Few	Yes	No	> 50%
Triton	Neptune	Yes	Yes	Few	No	Thin N_2	25%

Europa. Io may be geologically active, but Europa has an ocean. Recent observations of Europa by the Galileo mission add support to the idea that it is basically a small rocky core, surrounded by a deep, salty ocean, and topped with a thin, water-ice crust. The ocean is kept liquid by tidal interactions with Jupiter and nearby moons. Figure 4-3 shows surface features which strongly resemble ice floes on Earth.

Ganymede and Callisto. Ganymede and Callisto are relatively close in size, and we might expect that they would have similar histories. But Ganymede has been geologically active, and shows fault-like regions, and long, parallel ridges thousands of kilometers long. The entire surface is cratered, and indicates that the surface stopped evolving after only about one billion years. Callisto, on the other hand, shows absolutely no evidence of ever having been geologically active.

84　CHAPTER 4　Jovian Planets and Their Satellites

Fig. 4-2. Picture of Io, with Prometheus on limb. (Courtesy of NASA/JPL.)

The surface of Callisto looks much like the surface of the Moon, completely covered by craters.

The difference may be due to a slight difference in composition. Ganymede has about 5% more rock than Callisto. This rock would have inhibited convection a bit more strongly than ice. In addition, the rock would have contributed some radioactive heating to Ganymede. Both of these factors may have contributed to make Ganymede stay warm longer than Callisto.

CHAPTER 4 Jovian Planets and Their Satellites — 85

Fig. 4-3. Galileo image of the crust of Europa. (Courtesy of NASA.)

SATURN'S MOONS: TITAN

Titan is one of the largest moons in the solar system, and the only other object with a thick N_2 atmosphere like the Earth's. This atmosphere is thick with aerosols (particles suspended in air) which obscure the surface. Infrared images of the surface reveal bright and dark spots, which might be continents, and lakes or oceans of ethane, respectively.

NEPTUNE'S MOONS: TRITON

Triton is slightly smaller than Europa. It is the only major satellite with a retrograde orbit, which indicates that it is probably a captured object. Because of tidal forces, Triton's orbital speed is slowing, making it fall towards Neptune. Eventually, it will be torn apart by Neptune's gravity, and perhaps will form a ring system as spectacular as the one around Saturn.

Triton has a very thin N_2 atmosphere, relatively few impact craters, and other evidence of past geologic activity. Currently, Triton appears to be geologically active, with geysers of N_2 large enough to be seen by Voyager 2.

Solved Problems

4.6. Suppose one-third of the mass of a satellite was rock ($\rho = 3{,}500 \, \text{kg/m}^3$), and the rest was ice ($\rho = 900 \, \text{kg/m}^3$). What is the average density?

To find the average density, simply add up the components of the satellite, weighted by what fraction they make up:

$$\rho_{\text{avg}} = \frac{1}{3} \text{ rock} + \frac{2}{3} \text{ ice}$$

$$\rho_{\text{avg}} = \frac{1}{3}(3{,}500) + \frac{2}{3}(900)$$

$$\rho_{\text{avg}} = 1{,}770 \, \text{kg/m}^3$$

The average density of the satellite is $1{,}770 \, \text{kg/m}^3$.

4.7. Titan and the Moon have similar escape velocities. Why does Titan have an atmosphere, but the Moon does not?

Titan is farther from the Sun, so its surface temperature is much lower, about 100 K. With a temperature this low, an object of this size can retain its atmosphere.

A secondary consideration is that Titan is geologically active, due to tidal heating. Outgassing geysers, which add more N_2 to the atmosphere, are constantly renewing the atmosphere on Titan. The Moon, on the other hand, is geologically dead. The primordial atmosphere has been lost to space, and it has no means to replenish it.

4.8. Assume one of the newly discovered moons of Jupiter has a density of about $1{,}800 \, \text{kg/m}^3$. What is its approximate composition?

By comparing this density to the density of rock ($3{,}500 \, \text{kg/m}^3$), we know that the moon has very little heavy metals inside. But still, the density is higher than water, by about 50%, so the object is not completely water ice. Probably, it is about half water ice and half rock. This compares well with the composition of Titan, which has a density of $1{,}880 \, \text{kg/m}^3$.

4.9. A volcano on Io can throw material about 1,000 times higher than a similar volcano on Earth. Why?

There are two reasons for this. First, the gravity on Io is lower, so the material is not pulled down to the surface as strongly. Secondly, there is very little atmosphere on Io, so the air resistance is less.

4.10. Why is the Earth's Moon not kept geologically active by tidal heating?

In order for a satellite to experience tidal heating, it must change its orientation relative to the planet. Since the Moon is always in the same orientation relative to the Earth, there is no friction generating heat in the interior.

4.11. Why are old, large craters on Ganymede and Callisto much shallower than those on the Moon?

CHAPTER 4 Jovian Planets and Their Satellites

Ganymede and Callisto have shallower craters because they are icy worlds instead of rocky ones. The icy surface is less dense, and disperses more of the impact energy than the rocky surface does.

Rings

In addition to moons, the outer planets are also surrounded by ring systems. The most well known of these are the rings of Saturn, but Jupiter, Neptune, and Uranus also have ring systems. Rings occur inside the **Roche limit** of a planet. The Roche limit indicates how close to a planet an object can be before tidal forces overcome the object's own gravity. In other words, when an object comes too close to a planet, the difference in force between the near side and the far side is stronger than the force of gravity holding the object together, so it disintegrates. The Roche limit is generally about 2.5 times the planet's radius.

Rings are transitory objects. They form when a moon, asteroid, or other solid body is demolished by tidal forces, and the particles travel in an orbit around the planet. Some of these particles travel faster than others, and they move farther from the planet. The slow ones move closer, and the ring spreads out. The particles collide with each other, transferring energy back and forth, so that some fall inward, and some travel outward. Eventually, the whole system dissipates, with some particles added to the planet's mass, and some escaping the system entirely. This takes a few hundred million years, an astronomically short period of time.

The rings of the outer planets are very thin. When viewed edge-on, they disappear all together. Collisions play the major role in flattening the rings, and in keeping them flat. When a "south-bound" particle collides with a "north-bound" particle, the vertical parts of the velocities cancel out, and both particles wind up in a more "east–west" trajectory. Over time, the vertical motions of most particles are eliminated, and the rings become quite flat.

The rings of Saturn, however, show evidence that the rings are not completely flat. In some portions of the rings, **spokes** appear. These spokes form very quickly, and appear as dark features across the rings. The spokes are probably the result of the collision of a small meteoroid with the rings, which vaporizes, producing charged dust. This charged dust is levitated by the magnetic field of Saturn, and casts a shadow on the rings, which appears as a spoke.

All of the rings of the outer planets are made of many much thinner ringlets. A few of these ringlets are kept in place by **shepherding satellites**: small moons that push and pull the even smaller ring particles so that they stay in the ring.

Many of the ringlets appear brighter when backlit. This means that they are made of very small particles, which prefer to scatter light forward rather than backward. Larger dust particles prefer to scatter light backward, and so are brighter when viewed from the directly illuminated side.

The rings of Uranus were discovered in 1977 during an occultation of a star. As a star passed behind the rings, its light was dimmed repeatedly. Then it passed behind the planet. As the star emerged on the opposite side of the planet, and passed behind the rings again, the light was dimmed again, in a reverse of the pattern on the opposite side. Voyager 2 observed the rings directly, and found they contain a great deal of fine dust. This means they must have formed quite recently (within the last 1,000 years or so), because small dust particles have easily disturbed orbits. Perhaps these rings are replenished by meteoroid impacts on tiny satellites.

Solved Problems

4.12. If a comet came close enough to pass through the Roche limit, but was traveling fast enough to escape again, what would happen to it?

The comet would break up into smaller fragments due to tidal forces. Fragments would continue to orbit the Sun but in a new orbit.

4.13. The space shuttle orbits within the Roche limit of the Earth, yet is not pulled apart by tidal forces. Why not?

The Roche limit is the radius within which tidal forces overcome an object's gravity. But the space shuttle is not held together by gravity, it is held together by bolts and welding (it has "mechanical strength"), which are much stronger than gravity.

4.14. Why are rings thin? Are any other systems that you have learned about similar to the rings?

Collisions dominate ring processes. In this case, the collisions between particles which move in and out of the rings cancel out the velocities in those directions, and the particles are left with velocity only in the plane of the ring.

During the formation of the solar system, a thin disk formed. This disk was similar to the rings, because it orbited a larger body, was composed of bodies which were tiny in comparison, and was thin compared with its radius.

4.15. Why are some rings bright from one side, but dark from the other?

The brightness of a ring, and how it scatters light, depends on the size of the particles. When the particles are small, they scatter light (reflect it in all directions). The rings will appear brightest when backlit. When the particles are large, the light will generally be reflected back in the direction from which it came, as from a mirror. These rings appear brightest when "front-lit."

4.16. We see rings around planets, but theory says they don't live very long. What does that tell you about the rings we see today?

The rings that we see today must either have formed very recently, perhaps by collisions of small satellites, or they must be constantly replenished. It is likely that Saturn's rings, which

CHAPTER 4 Jovian Planets and Their Satellites

are composed of larger (meteor-sized) particles, are recently formed, since the particles haven't collided enough times to be pulverized. Other ring systems, however, probably are replenished occasionally, since the particles tend to be quite small.

4.17. Explain an occultation and how it could be used to map the density of rings. When might this method fail to detect a ring that is actually present?

An occultation occurs when a distant star passes behind an object, and the starlight gets "occluded," or dimmed. This can be used to map out the density structure of rings because the pattern of dimming and brightening will indicate where the rings are dense and where they are thin. This method will fail to detect rings made of very small dust particles, which will scatter the light forward, so that it is not substantially dimmed.

Supplementary Problems

4.18. What is the escape velocity from Titan ($M = 1.34 \times 10^{23}$ kg, $R = 2.575 \times 10^6$ m)?

Ans. 2.6 km/s

4.19. Where is the Roche limit of Saturn?

Ans. 151,000 m from the center of Saturn

4.20. Why are the current and past geological activity on Titan unknown?

Ans. Because it has an opaque atmosphere

4.21. Given that the orbital period of Io is 1.77 days, and the semi-major axis of its orbit is 4.22×10^8 m, calculate the mass of Jupiter.

Ans. 1.9×10^{27} kg

4.22. What is the gravitational attraction between Jupiter and Callisto ($M = 1.07 \times 10^{23}$ kg, $d = 1.88 \times 10^9$ km)?

Ans. 3.8×10^{21} N

4.23. What physical principle allows us to conclude that Triton must be a captured object?

Ans. Conservation of angular momentum

CHAPTER 5

Debris

Comets

Comets are balls of dirty ice from the outer solar system that follow elliptical orbits with high eccentricities, so that they are near to the Sun for only a small portion of their lives. As a comet comes near to the Sun at perihelion, the outer layers heat up and turn to gas, causing a coma (halo) and a tail to form. Very close to the Sun, the tail of a comet splits into two pieces, an **ion** or **plasma tail** and a **dust tail**. While both tails point away from the Sun, the dust tail curves "back" along the orbit, while the plasma tail is swept straight away from the Sun by the solar wind. These tails can be as long as 1 AU, making comets the largest objects in the solar system. However, comet tails are extremely diffuse; comet tails are more perfect vacuums than any we can make on Earth. The entire mass of a comet is less than 1 billionth the mass of the Earth.

The **nucleus** of a comet is a few kilometers across, and contains lots of water ice and carbon dioxide ice. This nucleus is surrounded by the **coma**—this is the "head" of the comet. The coma can be over 1 million km across. The coma shines both by reflected sunlight, and by the transitions of excited atoms and molecules in the gas (Fig. 5-1).

Comets can be divided into two types—long-period and short-period comets. This distinction is not quite as arbitrary as it sounds, since there are two different reservoirs for comets in the solar system. The long-period comets come from the **Oort cloud**, a swarm of comets 50,000–100,000 AU from the Sun. These comets have been in the Oort cloud since the solar system formed, and contain material that has remained the same since before the Sun formed. The Oort cloud is approximately spherical in shape, although there is probably a denser region near the plane of the solar system. An ice ball leaves the Oort cloud to become a comet when a star passes nearby (within 3 light years), and changes the ice ball's orbit. The passage of a star slows the ice ball, so that it no longer has enough energy to maintain its orbit. These objects fall into long elliptical orbits around the Sun. It is rare for such an event to happen; about 10 stars per million years pass

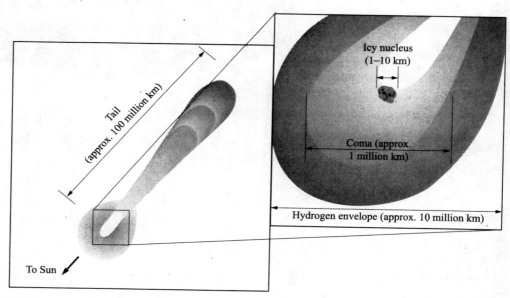

Fig. 5-1. Anatomy of a comet.

close enough to change the orbits in the Oort cloud. Each star may affect several ice balls, however. There are probably trillions of icy balls in the Oort cloud.

Short-period comets have periods less than 200 years, and originate in the Kuiper belt. The Kuiper belt is located just outside the orbit of Neptune, between about 30 and 50 AU from the Sun. These comets are distributed in a flat ring on the ecliptic. Extrapolating from known Kuiper belt objects indicates that there are probably about 70,000 comets in the Kuiper belt larger than 100 km across.

Each time a comet passes near the Sun, it sheds some of its mass, which remains in the orbital path. Eventually, the comet disintegrates entirely, unless, of course, it runs into the Sun, a planet, or receives a gravitational "assist" out of the solar system during one of its orbits.

Solved Problems

5.1. How could you find out how much of a comet's light is reflected, and how much is emitted?

The reflected sunlight will have the same spectrum as the Sun. It will be a blackbody, with Sun-like emission and absorption lines. Other emission lines may be present, and all of these will be from excited molecules and atoms in the comet itself.

CHAPTER 5 Debris

5.2. In what part of the orbit can a comet travel with the plasma tail "in front"? Draw a diagram to explain.

Just after the comet passes the perihelion of its orbit (closest approach to the Sun), the plasma tail will swing around because it always points away from the Sun. It is in this part of the orbit that the plasma tail can lead (Fig. 5-2).

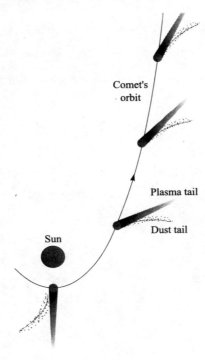

Fig. 5-2. Sometimes the plasma tail is in front of the comet.

5.3. Given that short-period comets come from just outside the orbit of Neptune, (a) how long is the period of a short-period comet? (b) Halley's comet has a period of 76 years. Is it a long- or short-period comet?

(a) If the comet comes from just outside the orbit of Neptune, at 30 AU from the Sun, and comes to within about 1 AU of the Sun, then the semi-major axis is *approximately* 15.5 AU. Using Kepler's third law,

$$P^2 = a^3$$
$$P = \sqrt{15.5^3}$$
$$P = 61 \text{ years}$$

(b) Comets from the Kuiper belt have periods less than 200 years. With a period of 76 years, Halley's comet is definitely a short-period comet.

5.4. From the orientation of an orbit, is it possible to determine whether a comet comes from the Oort cloud or the Kuiper belt?

Sometimes. If the orbit is highly inclined to the ecliptic, so that the comet cannot come from the Kuiper belt, then it is probably a long-period comet from the Oort cloud. However, if the orbit is on the ecliptic, it could have come from either source, and further observations are necessary (of the speed, for example, or the shape of the orbit) to determine the source of the comet.

5.5 Why do the plasma and dust tails usually point in different directions?

The plasma tail extends almost directly away from the Sun, while the dust tail is curved. Usually, the comet is not moving on a path which points directly toward the Sun. This is the only time that the two tails will point in exactly the same direction (Fig. 5-3).

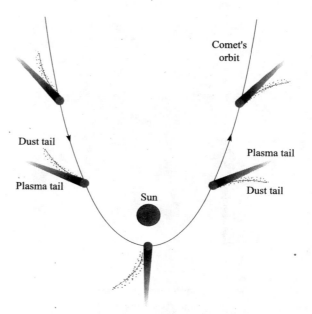

Fig. 5-3. The directions of the two comet tails. Note that far from perihelion, the comet will not have tails.

5.6. Which would hit the Earth at higher velocity: a prograde comet (one that orbits the Sun in the same direction as the Earth) or a retrograde comet (one that orbits oppositely)?

A retrograde comet would impact at higher velocity because the Earth and the comet would be heading towards each other. The prograde comet would be traveling in the same direction, so the relative velocity would be lower than the actual velocity. As an analogy, imagine two cars traveling at 60 mph and at 62 mph. In a head-on collision, the impact velocity is $60 + 62 = 122$ mph. If the faster car hits from behind, then the impact velocity is only $62 - 60$ mph or 2 mph.

CHAPTER 5 Debris

Meteorites

When a small rock or bit of dust is floating in space, it is a **meteoroid**. As it falls through the atmosphere of the Earth, it produces a bright streak of light, and is called a **meteor**. When it actually makes it to the surface of a planet or moon, we call the rock a **meteorite**. The brightest meteors are called **fireballs**. Sometimes these are as bright as the full moon. **Micrometeorites** are meteorites that are as small as sand grains. These are so small that the atmosphere slows them without heating them, and they drift to the surface of the planet. About 100 tons of micrometeorites accumulate on Earth every day. On the Moon, however, there is no atmosphere, and the micrometeorites are not slowed before they hit the surface. This is the main erosion process on the Moon, and the major contributor to the regolith.

There are three basic types of meteorites: iron, stony, and stony-iron. **Iron** meteorites are the easiest to recognize. They are overly heavy for their size, because they have a high proportion of iron. The so-called Widmanstätten patterns (Fig. 5-4) are observed in polished and etched slices of these meteorites and provide evidence that these meteorites come from planetesimal-sized chunks of rock. The size of the crystals indicates how slowly the rock cooled. If the meteorite was formed at its current size, it would have cooled quickly, and these crystals would not have formed. If the meteorite had formed in a large (~100 km) object, it would have been under higher pressures, and cooled much more slowly. These patterns imply that the meteorite cooled over millions of years, which is consistent with an object the size of a planetesimal.

Fig. 5-4. Widmanstätten patterns in an iron meteorite.

Stony meteorites resemble ordinary rocks. Consequently, stony meteorites are much less likely to be found, even though they are much more common than iron meteorites (95% of the meteorites that fall to Earth are stony meteorites). Stony meteorites have about the same density as ordinary rock, and hence are more difficult to find. Most of these stony meteorites are found in places like Antarctica, or the Sahara desert, where there are few ordinary rocks on the surface. Most stony meteorites contain rounded particles imbedded in the rest of the rock (Fig. 5-5). These lumps are called **chondrules** and the entire stony meteorite is then called a **chondrite**. **Carbonaceous chondrites** are a special kind of chondrite that contain high levels of carbon and often contain amino acids, the building blocks of proteins.

Fig. 5-5. A meterorite with chondrules. (Courtesy of New England Meteoritical Services.)

Stony-iron meteorites, a hybrid in which pieces of metal are embedded in ordinary silicate rock, are less than 1% of the total number of meteorites that fall to Earth.

The majority of meteorites probably come from the asteroid belt. Asteroids are large enough to have held the heat of the early solar system for millions of years. This allowed them to differentiate, so that the iron fell to the center, surrounded by a thin stony-iron layer, and enveloped in a thick stone "crust." When two such objects collide, the fragments consist of lots of stony meteoroids, fewer iron meteoroids, and a very small number of stony-iron meteoroids. These fragments spray away from the collision site, and a few of them eventually find their way to planets. Other sources of meteorites are comets, the Moon, and Mars.

Meteor showers are caused by a different phenomenon. As comets disintegrate, they leave behind in their orbits dust and larger debris ranging in size from millimeters to centimeters. Some of the comet orbits intersect the Earth's orbit, so that the Earth passes through them once each year. The infall of the larger debris causes meteor showers, when the dust particles stream through the atmosphere more often than normal. These particles are not large enough, in general, to result

CHAPTER 5 Debris

in a meteorite. The meteors produced in a meteor shower all appear to come from the same point in the sky. This point is called the **radiant** (because the meteors radiate away from it). The location of the radiant gives the meteor shower its name. For example, the radiant of the Leonids is in the constellation Leo. The radial pattern of the meteors is a product of perspective, because the Earth is passing through the meteoroid swarm. Figure 5-6 shows a time-lapse picture of a meteor shower.

Fig. 5-6. A time-lapse photograph of a meteor shower. (Image courtesy of NASA.)

Solved Problems

5.7. Why do meteorite hunters search for fireballs, but ignore meteor showers?

Meteors in meteor showers are usually produced by small particles that generally burn up in the atmosphere at altitudes of 30–100 km. (The smallest ones produce no bright streak at all.) Fireballs are caused by larger objects, which may not completely burn up in the atmosphere.

The potential for finding a rock at the end of a fireball trail is much higher than the potential for finding a rock at the end of a meteor trail which is part of a meteor shower.

5.8. Why are carbonaceous chondrites fundamentally important to the question of the existence of extraterrestrial life?

Carbonaceous chondrites contain amino acids, commonly known as the building blocks of life. If these amino acids are extraterrestrial in origin, they indicate that there are amino acids in space. This means that life does not need to originate from scratch on every planet or moon independently, but could be assisted from the impact of these meteoroids.

5.9. What is the kinetic energy (KE) of a 1,000 kg rock traveling at 30 km/s? Express your answer in megatons of TNT (the usual unit of measure for nuclear warheads—1 megaton of TNT $= 4 \times 10^9$ joules).

Convert the velocity from 30 km/s to 30,000 m/s. Use the kinetic energy equation

$$KE = \tfrac{1}{2}mv^2$$
$$KE = \tfrac{1}{2}(1,000)(30,000)^2 \text{ kg} \cdot \text{m}^2/\text{s}^2$$
$$KE = 9 \times 10^{11} \text{ joules}$$
$$KE = \frac{9 \times 10^{11} \text{ J}}{4 \times 10^9 \text{ J/Mton}}$$
$$KE = 225 \text{ megatons of TNT}$$

There is equivalent kinetic energy in a rock of this size to a 225-megaton warhead.

5.10. Why are meteor showers predictable?

Meteor showers are caused when the Earth intercepts the dust left behind in the orbit of a comet. Since the Earth is at the same place in its orbit on the same day every year, it intercepts the orbit of a comet on the same day every year.

5.11. Draw a diagram of a differentiated asteroid (Fig. 5-7).

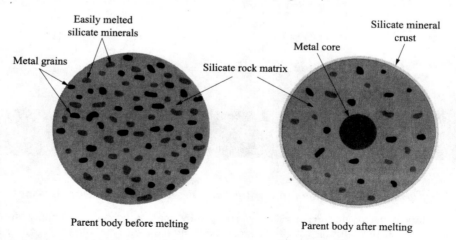

Fig. 5-7. A differentiated asteroid.

CHAPTER 5 Debris

5.12. When observations are skewed in favor of objects which are easily observed, this is called a "selection effect." Explain how the distribution of known meteorites is biased by a selection effect.

On Earth, the easiest meteorites to find are the iron meteorites, because they are easily distinguished by their weight. Stony meteorites are more common, but difficult to find because they closely resemble ordinary rocks. Only in places where terrestrial rocks are scarce (Antarctica, the Sahara desert), are stony meteorites easily distinguished from terrestrial rocks. This makes it far more likely that any *identified* meteorite is an iron meteorite, independent of the actual numbers of iron versus stony meteorites that land on the planet.

5.13. Suppose an iron meteorite has no Widmanstätten patterns. What does this indicate about the formation of the meteoroid's parent body?

The presence of Widmanstätten patterns indicates that the parent body cooled very slowly. If there are no Widmanstätten patterns, the parent body of this rock must have cooled quite rapidly—too rapidly for these patterns to form. This in turn implies that the parent body must have been small, so that it gave up its heat quickly.

5.14. The Leonid meteor shower lasts about 2 days. The Earth moves 2.5 million km each day. How thick is the belt of meteoroids that causes the Leonids?

The Earth moves 5 million km in 2 days, so the belt of meteoroids must be 5 million km thick.

Asteroids

There are about 1 million asteroids larger than 1 km in the solar system. The vast majority of these orbit the Sun between Mars and Jupiter. Over 8,000 of these have been individually cataloged and named, and have well-determined orbits. Although it is common to depict the asteroid belt as a dense region, asteroids are actually quite well separated, rarely approaching within 1 million km of one another. (A few asteroids have moons of their own: these are certainly the exception to the rule.) All together, the asteroid belt contains about 0.1% of the mass of the Earth.

The asteroids are not uniformly distributed throughout the asteroid belt. The so-called Kirkwood gaps are regions avoided by asteroid orbits. If an asteroid orbits at a radius such that its period is 1/2, 1/3, 1/4, etc., of Jupiter's period, then after 2, 3, 4, etc., orbits, respectively, the asteroid will meet Jupiter in the same place, and get a gravitational tug towards the same direction in space. Because these tugs are not random, but occur over and over at the same location, the effect adds up, and the asteroid is pulled out of that orbit. This leaves gaps, where few asteroids exist. The asteroids in these gaps are generally in transit, either inward or outward.

In addition to the asteroids in the belt, some asteroids share Jupiter's orbit. Asteroids in this special group are called Trojan asteroids, and they orbit about 60° ahead or behind Jupiter. Their orbits are stabilized by the combined gravity of Jupiter and the Sun. Over 150 of these are known; the largest is about 300 km in size.

Finally, some asteroids are "Earth-crossing" and are potential impactors. These asteroids come from three different groups—the Apollo, Aten, and Amor asteroids. Most of these are small, less than 40 km across, and so they are difficult to find in the sky. About 500 are known. Most of these will strike the Earth some time over the next 20–30 million years. Near-misses are common, and are often unpredicted. In 1990, an asteroid came closer to the Earth than the Moon. The asteroid was previously undiscovered, and was not noticed until after it had safely passed the Earth.

Asteroids do not emit visible light, they only reflect it. Astronomers determine the compositions of asteroids by comparing the spectrum of the light reflected by the asteroid and the spectrum of the Sun. Absorption lines that are present in the asteroid's spectrum, but not in the solar spectrum, must be due to elements or minerals in the asteroid. Asteroids are classified in three major groups: carbonaceous (C), silicate (S), and metallic (M). Most asteroids are C-type asteroids, with very low albedo and no strong absorption lines. The rest are mainly S type, with an absorption feature due to a silicate mineral, olivine.

The amount of light reflected from an asteroid towards the Earth changes as the asteroid tumbles through space. We can use this information to determine how quickly the asteroids rotate. About 500 asteroids have been studied well enough to determine their rotation periods, which are generally between 3 and 30 hours. Smaller asteroids have irregular shapes. The shape of small asteroids can be determined from analysis of the amount of radiation received over time (light curve), which fluctuates with a period equal to the period of the asteroid. More radiation is received when the asteroid is viewed perpendicular to the longest axis than when viewed perpendicular to the shorter axes.

Solved Problems

5.15. Why are comets icy, but asteroids are rocky?

Comets formed in the outer solar system. From the section on solar system formation (Chapter 3), condensation temperatures were lower there, so that ices could form. Asteroids, however, formed in the inner solar system, where condensation temperatures were higher, keeping ices from forming.

CHAPTER 5   Debris

5.16. Suppose that the Earth-crossing asteroids are evenly distributed, so that the impact rate is constant. How many years will pass, on average, between two impacts?

There are 500 Earth-crossing asteroids, all of which will impact in the next (approximately) 25 million years. This means that there are about

$$\frac{25 \times 10^6}{500} = 50,000$$

years between collisions.

5.17. Figure 5-8 shows a "light-curve" of an asteroid (a graph of the amount of light from the asteroid versus time). Label the portion of the curve where the longest axis is perpendicular to the line of sight. Label the portion of the curve where the shortest axis is perpendicular to the line of sight. What is the period of this asteroid?

The time between long-axis peaks is 705.75 − 705.55 days, or about 0.2 days. This is equivalent to 4.8 hours, but is only half of the period. So the period is about 9.6 hours.

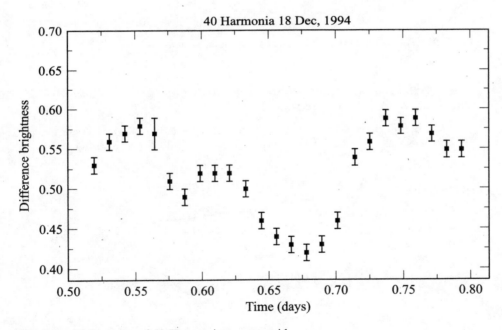

Fig. 5-8. Light curve of 40 Harmonia, an asteroid.

5.18. Sketch the distribution of asteroids in the solar system (Fig. 5-9).

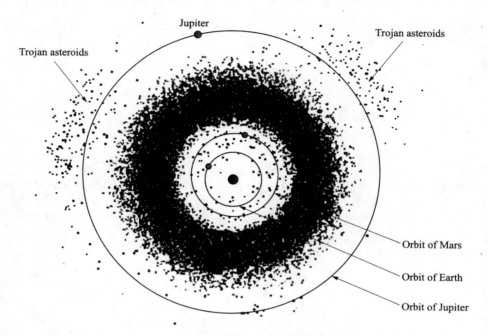

Fig. 5-9. The distribution of asteroids in the solar system.

5.19. Classify the following four spectra of asteroids into C type or S type (Fig. 5-10). How can you tell which are which?

C-type asteroids have no strong absorption lines. Asteroids A, C, and D are C-type asteroids, whereas asteroid B is an S-type asteroid because it does have a strong absorption line.

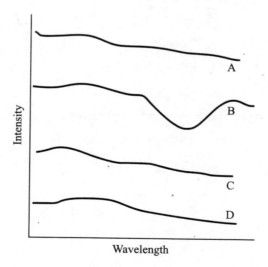

Fig. 5-10. Four spectra of asteroids—three C type and one S type.

CHAPTER 5 Debris

Pluto and Charon

Pluto is the smallest of the planets and fits neither the Jovian nor the terrestrial class. Pluto's orbit is highly inclined, about 17°, to the ecliptic, and has the most eccentric orbit of all the solar system planets. At its farthest, Pluto is 49.3 AU from the Sun; at its closest, 29.7 AU, it actually crosses inside the orbit of Neptune. This has led astronomers to believe that perhaps Pluto was an escaped Neptunian moon. Pluto's density is consistent with a composition of rock and ice. The thin atmosphere consists of nitrogen, with some carbon monoxide and methane. The surface temperature is about 40 K.

Pluto has a satellite, Charon, discovered in 1978. The tidal interaction between the two objects completely synchronized the motion of the pair, so that the two objects always show the same side to each other. One day on Pluto equals 1 day on Charon equals 1 synodic period. Astronomers now believe that both Pluto and Charon are Kuiper belt objects that were gravitationally perturbed (probably by Neptune), and drifted to the domain of the planets. Figure 5-11 shows Pluto and Charon superimposed on a map of the Earth in scale. The orbital and physical properties of Pluto and Charon are listed in Table 5-1.

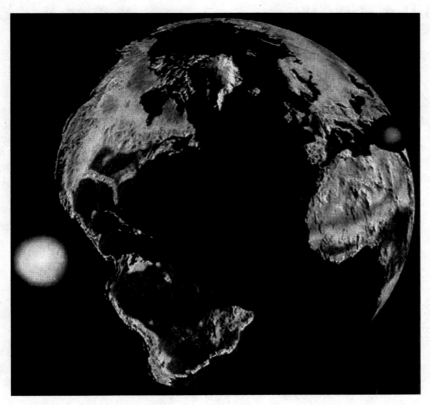

Fig. 5-11. Pluto and Charon superimposed on a map of the Earth. (Composite of NASA/STScI images.)

Table 5-1. Orbital and physical properties of Pluto and Charon

Property	Pluto	Charon
Mass (kg)	1.26×10^{22}	1.7×10^{21}
Radius (km)	1,150	~ 600
Mean orbit radius (km)	5.91×10^9	19.6×10^3 (from Pluto)
Orbital period	249 years	6.4 days
Orbital inclination	17	
Orbital eccentricity	0.25	
Rotation periods (days)	6.4	6.4
Tilt of axis	122.5	

Solved Problems

5.20. Explain how the inclination of Pluto's orbit makes it impossible for Pluto to collide with Neptune, even though the orbits cross.

There are two reasons for this. First, the plots of orbits as usually shown are only two-dimensional. The statement that they intersect, or cross, is misleading. The orbits cross at a location when Pluto is far above the plane of the solar system. This makes it impossible for Pluto and Neptune to ever collide. In addition, Pluto and Neptune are in resonance, much like the Trojan asteroids, so that they remain always separated by the same amount at the same points in their orbits. They will always repeat the same pattern while orbiting the Sun, which does not include a collision.

5.21. What is Pluto's orbital period?

Pluto's major axis is $50 + 30 = 80$ AU. The semi-major axis, then, is about 40 AU. Since Pluto orbits the Sun, we can use Kepler's third law,

$$P^2 = a^3$$
$$P = \sqrt{a^3}$$
$$P = \sqrt{(40)^3}$$
$$P = 253 \text{ years.}$$

The accepted orbital period of Pluto is 249 years. The 4-year discrepancy is largely due to the rounding in the distance from the Sun to Pluto at perihelion and aphelion, which each have only one significant digit.

CHAPTER 5 Debris

5.22. Draw a sketch of Pluto and Charon at several times over one Plutonian day (Fig. 5-12). Be sure to identify particular locations on each body, so that the rotations are clear.

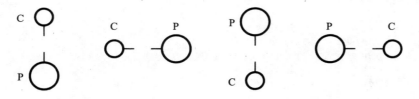

Fig. 5-12. Pluto and Charon over one Plutonian day.

Supplementary Problems

5.23. How long is the period of an average long-period comet?

Ans. 11 million years

5.24. Are the meteoroids in this year's Leonid shower located close (in space) to the ones from last year's meteor shower?

Ans. No, because the particles are orbiting too

5.25. Suppose a really spectacular comet approaches the Earth, with a coma of 1.5 million km in diameter. At closest approach, it is 0.9 AU from the Earth. What is the angular size of the coma?

Ans. $0.6°$

5.26. Suppose this same spectacular comet has a tail 1.3 AU long. What is the angular length of the tail?

Ans. $83°$, nearly halfway across the visible sky

5.27. What is the angular size of a 1-km diameter asteroid in the asteroid belt (at 3 AU)?

Ans. $5 \times 10^{-4}{''}$

5.28. The peak blackbody temperature of an asteroid is in the infrared. Why are they usually observed in the visible?

Ans. Because the majority of the light from an asteroid is reflected sunlight

5.29. The mass of Pluto is 0.0020 times the mass of the Earth. What is the mass of Pluto in kg?

Ans. 1.19×10^{22} kg

5.30. Charon orbits Pluto at a distance of 19,600 km. In Fig. 5-11, are Pluto and Charon shown at their greatest separation?

Ans. No

5.31. What is the orbital period of Charon (Charon's mass is 1.7×10^{21} kg)?

Ans. 6.47 days

5.32. What percentage of the asteroids larger than 1 km have been cataloged?

Ans. Approximately 1%

CHAPTER 6

The Interstellar Medium and Star Formation

The Interstellar Medium

The interstellar medium (ISM) is the dust and gas between the stars. The interstellar medium is seen as the dark dust lanes in the Milky Way (or in other galaxies), or by its effects on starlight: reddening and extinction. It is also observed more directly as reflection or emission nebulae.

Approximately 20% of the Galaxy's mass is ISM. The ISM absorbs visible light, but at the same time emits radio waves or infrared radiation. So, while we can't make an accurate map of the distant Galaxy in visible light, we can see distant pockets of interstellar dust and gas quite easily.

Ninety-nine percent of the ISM is gas, and only 1% of the mass is dust. Even in the nebulae, which have fairly high densities (Table 6-1), the density is lower than in the best vacuums that we can achieve in a laboratory.

GIANT MOLECULAR CLOUDS

Much of the mass in the interstellar medium is grouped into clumps called giant molecular clouds. These clumps are denser than the surrounding medium, and cool enough that they can contain molecular hydrogen (and trace amounts of other molecules). A typical giant molecular cloud is about 10 pc across, and contains about 1×10^6 solar masses. There are thousands of giant molecular clouds in the Milky Way Galaxy. The closest one is the Orion Nebula, shown in Fig. 6-1. The Orion Nebula is 450 pc away.

Table 6-1. Properties of components of the interstellar medium. Values are approximate, and are meant to be used only as a guide

Type of nebula	Density (g/cm^3)	Temperature (K)	Typical lifetime (yrs)	Typical size (pc)	Composition
HII region	10^{-25}–10^{-17}	10,000	10 million	few–100	Mostly hydrogen gas
Giant molecular cloud: cool clumps	10^{-20}–10^{-18}	10	Billions	0.1	Hydrogen, molecular gas, dust
Giant molecular cloud: hot clumps	10^{-25}–10^{-17}	30–100	Millions	0.1–3	Hydrogen, molecular gas
Reflection nebulae	10^{-25}–10^{-17}	<1,000	Millions–billions	<1–10	Dusty gas
Emission nebulae	10^{-25}–10^{-17}	1,000–10,000	Few thousand–100,000	0.01–a few	Atomic and molecular gas
Dark nebulae	10^{-25}–10^{-17}	<1,000	Millions–billions	<1–10	Dusty gas
Diffuse interstellar gas	10^{-27}	7,000–10,000	Not applicable	Not applicable	Hydrogen

CHAPTER 6 The Interstellar Medium

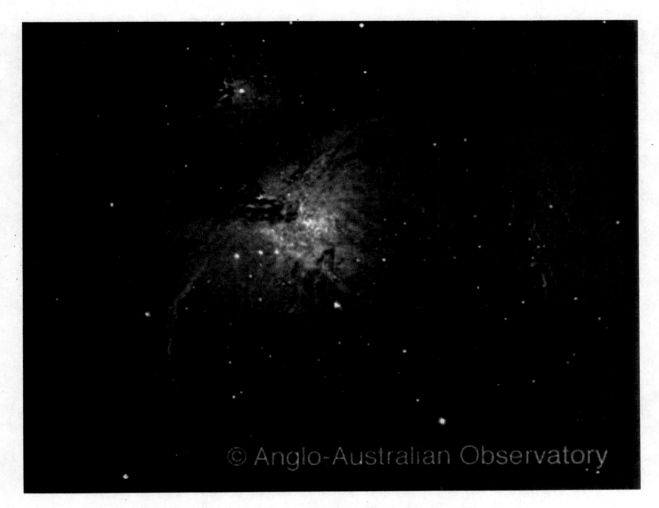

Fig. 6-1. The Orion Nebula in visible light. (Courtesy of the Anglo-Australian Observatory.)

INTERSTELLAR GAS

HII regions. These are usually parts of giant molecular clouds: specifically, the parts surrounding hot, young stars, such as O or B types (see Chapter 7). O and B stars emit UV light, which ionizes the hydrogen in the nebula. When the electrons recombine with the protons to form hydrogen, they emit photons at wavelengths characteristic of the hydrogen spectrum. The red color of HII regions is due to the 656 nm spectral line of hydrogen.

HII regions can be enormous, particularly if there is more than one O star producing UV radiation in the neighborhood. The O stars are massive, but have very short lifetimes. Consequently, the HII regions only exist for about 10 million years; when the O star exhausts its fuel, the HII region no longer has a source of ionizing photons and fades away.

Diffuse interstellar gas. The diffuse interstellar gas occupies the regions between the nebulae. It is extremely low density (therefore "diffuse"). Diffuse interstellar gas was discovered by observing the spectra of a binary star system. Astronomer J. Hartmann observed that there were really two sets of spectral lines from the binary (one from each star, see Problem 6.8), which shifted back and forth as the stars orbited each other. In addition to the shifting lines, there were spectral lines that did not shift. The fixed lines were narrow and were identified as lines of calcium in the intervening medium. Many other atoms have been found in the interstellar medium since the discovery of calcium.

In particular, neutral hydrogen has been identified by observing the 21 cm line. The 21 cm line results from a spin transition of the electron in the hydrogen atom. The electron and the proton each have a property known as the spin. When the spins of the two particles are parallel, they have more energy than when the spins are anti-parallel. The energy difference is very small, so that the photon emitted when the electron flips its spin is very low energy. That means that it has a long wavelength, 21 cm, and is observable as a radio wave.

Molecules have also been detected in the diffuse interstellar gas. The first molecules discovered were CH, CN, and CN^+, followed by H_2O (water) and NH_3 (ammonia). Currently, there are more than 80 known molecules in the diffuse interstellar gas, making up 0.002% of the interstellar gas. Some of these molecules are very complex, and contain as many as 16 atoms. Even amino acids have been found. Molecules are fragile, and are easily destroyed by UV radiation. For the most part, molecules only survive inside clouds, where they are protected by dust particles.

INTERSTELLAR DUST

There are three ways to detect the interstellar dust: infrared radiation from the warmed particles, extinction of visible light (dark patches), or reflection nebulae.

1. Dust is warmed by UV radiation, and therefore emits a blackbody spectrum that peaks in the infrared. Figure 6-2 shows the Milky Way Galaxy in the optical and the infrared. Dust emits in the infrared but blocks visible light. Therefore, the infrared image is bright whereas the optical image is dark, and vice versa.

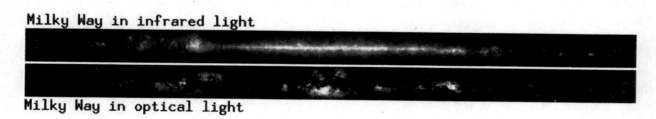

Fig. 6-2. Milky Way in optical and infrared light. (Courtesy of NASA.)

CHAPTER 6   The Interstellar Medium

2. Dark nebulae: the dark patches in the Milky Way and dark clouds observed in other places (such as the Horsehead nebula). These dark nebulae are often called dust clouds even though they consist mostly of gas. These clouds range in size from less than 1 pc to more than 10 pc. Most nebulae absorb 75% of the starlight from background stars—a few absorb more, nearly 95% in some cases.
3. Reflection nebulae: these shine by scattered starlight, in much the same way that the blue sky shines by scattered sunlight. These nebulae are always bluish in color, and so are easily distinguished from HII regions that are always reddish in color. The line spectrum of these nebulae resembles the spectrum of the stars whose light they scatter.

Diffuse interstellar dust.   Analogous to the diffuse interstellar gas is the diffuse interstellar dust. This dust is difficult to detect, and can only be found by looking at very distant clusters and comparing them to nearby clusters. The distant clusters will appear redder and fainter than they should at their distances. This is the result of two phenomena caused by the diffuse interstellar dust: extinction and reddening.

Extinction. On average, the diffuse interstellar dust dims starlight by one magnitude (= 2.51 times) per 1,000 pc in the plane of the Milky Way.

From a star at a distance of	Percent of light lost due to extinction
1,000 pc	60
2,000 pc	84
5,000 pc	99
10,000 pc	99.99

The Milky Way is about 60,000 pc across, so most of the starlight from distant stars has been extincted before reaching the Earth. This extinction is caused by both scattering and absorption. Determining which effect is more important is impossible without a detailed knowledge of both the composition and the size of the dust grains. As most of the interstellar dust is in the plane of the Milky Way, we can only see the galaxies that are "above" or "below" the Milky Way. On average, the Milky Way is about 2,000 pc thick.

Reddening. Dust scatters blue light more effectively than red light. As a result stars appear redder than they would if there were no dust in the line of sight. From a distance of 3,000 pc through the disk, only 2.5% of the blue light will reach us, whereas 6% of the red light will get through. The scattering efficiency decreases with decreasing frequency of light. This means that we can see deeply into the Milky Way using radio waves or infrared radiation, but not in ultraviolet radiation.

CHAPTER 6 The Interstellar Medium

Solved Problems

6.1. What is the peak wavelength of the emission from cool (100 K) dust?

Recall Wien's law from Chapter 1, which relates the peak wavelength and the temperature:

$$\lambda_{max} = \frac{k}{T}$$
$$\lambda_{max} = \frac{2.9 \times 10^{-3} \, m \cdot K}{100 \, K}$$
$$\lambda_{max} = 2.9 \times 10^{-5} \, m$$

The peak wavelength of 100 K dust is about 30×10^{-6} m, or 30 microns. This is in the infrared part of the spectrum.

6.2. How do astronomers determine the chemistry of a cool dust cloud?

As in many other applications, astronomers determine the presence of atoms and molecules in dust clouds from their emission and absorption lines. Most of the molecular lines that it is possible to observe occur in the infrared, millimeter, or radio wavelengths. Astronomers observe dust clouds with these types of telescopes, and compare the lines observed with lines formed in laboratory experiments.

6.3. What limits the age of an HII region?

HII emission is produced by gas which has been excited by UV photons. For this gas to be excited, there must be stars in the vicinity which are also producing lots of UV photons. These stars are in general of the O and B type, and so are massive and have short lifetimes. Once these stars end their lives (in about 10 million years), the HII region fades, and becomes non-luminous.

6.4. Describe the difference between reddening and extinction.

Interstellar dust absorbs and scatters starlight, dimming it. This is called extinction. Reddening depends on wavelength. Blue wavelengths are removed by scattering more than red wavelengths. This makes objects appear redder than they would if there were no dust in the way.

6.5. Figure 6-3 shows a spectrum of a dust cloud around a star. How hot is it?

The peak of the blackbody emission in this figure occurs at approximately 25 microns (25×10^{-6} m). Using Wien's law,

$$\lambda_{max} = \frac{k}{T}$$

or

$$T = \frac{k}{\lambda_{max}}$$
$$T = \frac{0.0029 \, m \cdot K}{25 \times 10^{-6} \, m}$$
$$T = 120 \, K$$

The dust cloud is 120 K.

CHAPTER 6 The Interstellar Medium

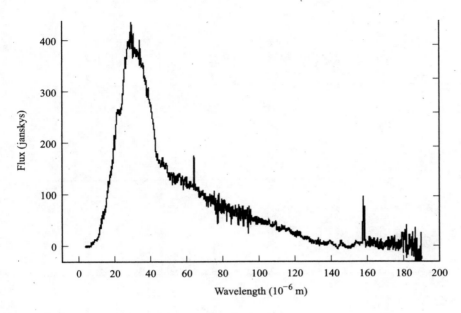

Fig. 6-3. A spectrum of circumstellar dust cloud.

6.6. How can you determine the difference between an emission and a reflection nebula?

There are two ways to determine the difference between an emission and a reflection nebula. The first method is to simply look at the color of the nebula. Since nebulae are made primarily of hydrogen, which has red emission lines, emission nebulae tend to be red. Reflection nebulae, however, shine by scattering starlight, and will be the color of the stars nearest them, usually much bluer than the emission nebulae. The second method is to look more closely at the spectra of the objects. A few sharp emission lines will dominate spectra of emission nebulae, while, in general, spectra of reflection nebulae look like the spectra of the stars whose light they are scattering.

6.7. What fraction of the Galaxy's mass is interstellar dust?

The interstellar medium is approximately 20% of the Galactic mass. Only 1% of this is dust. Therefore, $0.01 \cdot 0.2 = 0.002$ or 0.2% of the Galaxy's mass is interstellar dust. This is an extremely small component, yet it dominates our ability to see through the disk of the Galaxy.

6.8. Figure 6-4 shows several spectra of a binary star system. Label each line A, B, or C, depending on whether it came from star A, star B, or the interstellar medium.

6.9. How many solar-type stars could be made from a giant molecular cloud of diameter $D = 10 \, \text{pc}$ and density $\rho = 1.6 \times 10^{-17} \, \text{kg/m}^3$? Assume the cloud is spherical, and about half of it is used to form stars.

The volume of the giant molecular cloud is

$$V = \frac{4}{3} \cdot \pi \cdot R^3$$

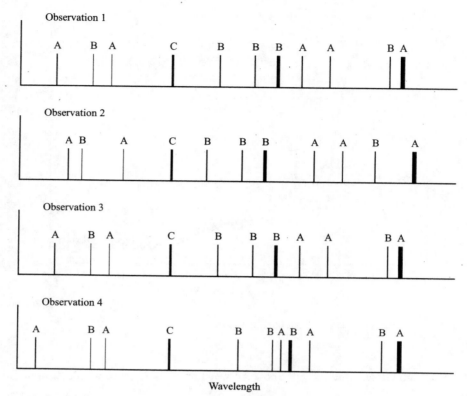

Fig. 6-4. Spectra of a binary star system, including interstellar lines, over several years.

The mass of the giant molecular cloud is found by multiplying the density by the volume,

$$M = \rho \cdot V$$

If only half of this mass forms stars, then we must divide by two to find the mass of all the stars formed. Substituting the first equation into the second gives

$$M = \frac{1}{2} \cdot \rho \cdot \frac{4}{3} \cdot \pi \cdot \left(\frac{D}{2}\right)^3$$

$$M = \frac{2 \cdot \pi}{3 \cdot 2^3} \cdot \rho \cdot D^3$$

$$M = 0.26 \cdot 1.6 \times 10^{-17} \text{ kg/m}^3 (10 \text{ pc})^3$$

$$M = 4.2 \times 10^{-15} \text{ kg/m}^3 \cdot \text{pc}^3$$

$$M = 4.2 \times 10^{-15} \text{ kg/m}^3 \cdot \left(\text{pc} \cdot \frac{3.08 \times 10^{16} \text{m}}{\text{pc}}\right)^3$$

$$M = 4.2 \times 10^{-15} \cdot (3.08 \times 10^{16})^3 \text{ kg}$$

$$M = 1.2 \times 10^{35} \text{ kg}$$

To find out how many solar mass stars could be produced, divide by the mass of the Sun (2×10^{30} kg) to find $M = 61,000 \, M_{\text{Sun}}$. About 60,000 stars with the same mass as the Sun could be formed from this average-sized molecular cloud.

CHAPTER 6   The Interstellar Medium

6.10. What is the difference in energy of the two spin states of the hydrogen atom?

Recall from Chapter 1 that the energy of a photon is given by

$$E = hf$$

and that the relationship between wavelength and frequency is

$$f = \frac{c}{\lambda}$$

Substituting the second equation into the first, and knowing that the wavelength of the photon is 21 cm gives an energy of

$$E = \frac{hc}{\lambda}$$
$$E = \frac{6.626 \times 10^{-34}\,\text{W} \cdot \text{s}^2 \cdot 3 \times 10^8\,\text{m/s}}{0.21\,\text{m}}$$
$$E = 9.4 \times 10^{-25}\,\text{W} \cdot \text{s}$$
$$E = 9.4 \times 10^{-25}\,\text{joules}$$

This is about 0.5×10^6 times smaller than the energy difference corresponding to the emission of visible light.

6.11. A radio spectrum of an interstellar cloud shows the 21 cm line shifted to 21.007 cm. Is the cloud approaching, receding, or remaining at the same distance from us? If it is traveling, what is its speed?

Because the wavelength increases, the emission is red-shifted, so the cloud must be moving away from us. To find its radial velocity, use the Doppler equation

$$\frac{v}{c} = \frac{\Delta\lambda}{\lambda}$$
$$v = \frac{\Delta\lambda}{\lambda} \cdot c$$
$$v = \frac{(21.007 - 21)}{21} \cdot 3 \times 10^8\,\text{m/s}$$
$$v = 100{,}000\,\text{m/s}$$
$$v = 100\,\text{km/s}$$

The cloud is moving away from us at 100 km/s.

6.12. Suppose that a cool clump of a giant molecular cloud has a size of 0.1 pc. Given the density range from Table 6-1, what is the possible range of masses contained in this clump?

The mass in a clump is given by the density times the volume,

CHAPTER 6 The Interstellar Medium

$$M = \rho \cdot V$$
$$M = \rho \cdot \frac{4}{3}\pi \cdot R^3$$
$$M = \frac{4}{3}\pi \cdot 10^{-20}\frac{\text{g}}{\text{cm}^3}(0.05\,\text{pc})^3$$
$$M = 5.2 \times 10^{-24}\frac{\text{g} \cdot \text{pc}^3}{\text{cm}^3}$$
$$M = 5.2 \times 10^{-24}\,\text{g} \cdot \left(\frac{\text{pc}}{\text{cm}} \cdot \frac{3 \times 10^{18}\,\text{cm}}{\text{pc}}\right)^3$$
$$M = 5.2 \times 10^{-24} \cdot (3 \times 10^{18})^3\,\text{g}$$
$$M = 1.4 \times 10^{32}\,\text{g}$$
$$M = 0.1 \times 10^{30}\,\text{kg}$$
$$M = 0.1 \text{ solar masses.}$$

The mass in a low-density clump of this size is 0.1 solar masses, not enough to form a star.
If the clump has the maximum density for a cool clump, the density increases by a factor of 100, but all else stays the same, so that the mass is

$$M = 100 \cdot (\text{cool clump mass})$$
$$M = 100 \cdot 0.1 \text{ solar masses}$$
$$M = 10 \text{ solar masses}$$

Therefore, the mass in a high-density clump could form approximately 10 solar systems, even though it is about the same volume as the low-density clump which could not form any solar mass systems.

6.13. What are the temperatures, in degrees Fahrenheit, of cool and warm clumps of giant molecular clouds? Do non-astronomers usually call these temperatures "cool" and "warm"?

To convert from degrees Kelvin to degrees Celsius, subtract 273.

Cool clumps: $-263°C$

Warm clumps: (-243) to $(-173)°C$

To convert from degrees Celsius to degrees Fahrenheit, multiply by 9/5, and add 32:
Cool clumps:

$$T(°F) = T(°C) \cdot 9/5 + 32$$
$$T(°F) = -263 \cdot 9/5 + 32$$
$$T(°F) = -441$$

Warm clumps:

$$T(°F) = T(°C) \cdot 9/5 + 32$$
for warmer warm clumps,
$$T(°F) = -173 \cdot 9/5 + 32$$
$$T(°F) = -279$$
for cooler warm clumps,
$$T(°F) = -243 \cdot 9/5 + 32$$
$$T(°F) = -405$$

These are all very cold.

CHAPTER 6 The Interstellar Medium

Star Formation

Clumps in giant molecular clouds are often sources of intense infrared radiation, the product of heated dust. If the cloud is collapsing, the dust heats up because as the dust falls in towards the center, the potential energy is released as heat. This idea of big, warm, collapsing clouds is at the center of star formation theory.

The mechanism that triggers the collapse into clumps is not well understood. Two mechanisms are commonly proposed. The first is that the cloud is simply turbulent to begin with, with some clumps and some thin parts, and then gravitational attraction and collisions make the clumps grow. The second is that winds from a supernova in the vicinity sweep up dust and gas, causing denser regions to form. Again, gravity does the majority of the work in making these clumps collapse.

For a clump to collapse, it must be large enough and dense enough so that the average velocity of a particle is less than the escape velocity.

Under the influence of gravity alone, clumps should take approximately 100,000 years to collapse. But we know from star counts that it must take much longer for stars to form; otherwise, there would be many more of them. The galactic magnetic field, which threads the cloud, slows the collapse. Charged particles, such as protons, have a very hard time crossing magnetic field lines. When a charged particle interacts with a magnetic field line, it begins to orbit the line, and can travel easily *along* the field line, but not *across* it (Fig. 6-5). Neutral particles (like neutrons) do not have this problem, and can cross the field lines to gather in the center of the clump. After a few million years, the gravitational force due to the neutral particles in the center overwhelms the magnetic field's resistance, and mass accumulates rapidly in the core. From this point to the formation of an actual star is only 100,000 years.

A **protostar** is a clump of a giant molecular cloud that is collapsing rapidly, but has not yet formed a star. Protostar is the stage from shortly before the gravity overcomes the magnetic field to the time that the star ignites. Recently, astronomers have been able to acquire direct images of some of these objects with the Hubble Space Telescope (Fig. 6-6). Several things happen during the protostar stage.

1. **Material falls onto the protostar**. As the material falls onto the star, it becomes denser and therefore more opaque. As objects become denser and more opaque, energy/photons require longer time to escape. As a result, the temperature of the protostar begins to rise rapidly.

2. **A disk forms**. Rotation causes the cloud to flatten as it collapses. Also, because of conservation of angular momentum, the rotational speed of the disk increases as the material spirals inward. These disks are made of warm dust that emits in the infrared.

    Friction in the disk causes the material of the inner part of the disk to spiral in and the material in the outer part of the disk to spiral out. These disks will eventually disappear because of:

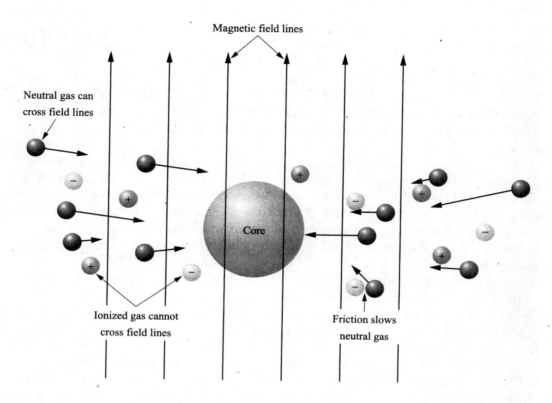

Fig. 6-5. Charged particles cannot cross magnetic field lines easily.

    (a) evaporation by a nearby object—if a bright star is nearby, it can evaporate the materials of the disk, and disperse it back to the interstellar medium;

    (b) wind dispersal—once the wind from the central star begins, it can transfer momentum to the disk particles, and disperse them back to the interstellar medium (see point 4 below); and

    (c) planet formation—planets may form from the disk; see Chapter 3 for more details on planet formation.

3. **Temperature and pressure increase.** Once the protostar becomes opaque, it can no longer radiate away the energy produced by the infalling material. This energy is trapped inside the protostar, and causes the temperature to rise. As the mass falls inward, the density increases. From the ideal gas law (Chapter 1), the result is an increase in the pressure. The pressure increase in the interior slows, but does not quite stop the collapse.

4. **A wind develops.** Astronomers are not quite sure why, but at this stage in the evolution, the protostar begins pushing mass away. This stops the infall of more mass. The winds produced are very massive, about 10^{-7} solar masses per year, and carry with them not only energy and mass but also angular momentum. If the Sun were losing mass this quickly, it

CHAPTER 6 The Interstellar Medium

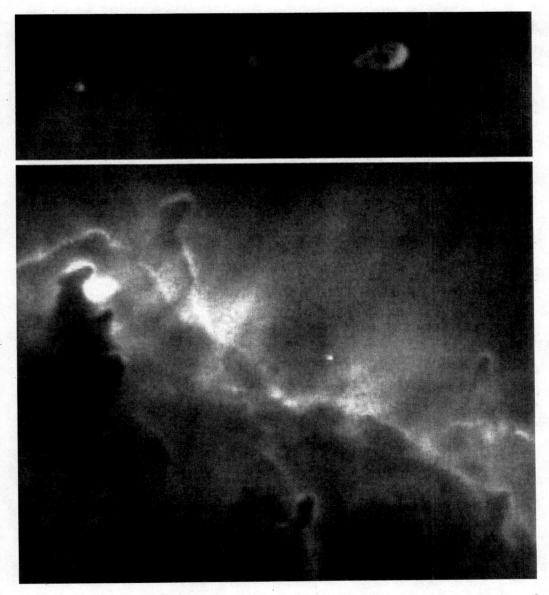

Fig. 6-6. An image of protostars in the Orion Nebula and in the Eagle Nebula. (Courtesy of STScI.)

would lose all of its mass in only 10 million years—an astronomically short period of time.

5. **Pressure and temperature increase further**. The pressure and the temperature continue to rise in the central portion of the protostar. Eventually hydrogen fusion begins, and the protostar moves onto the main sequence, and is finally a star.

CHAPTER 6 The Interstellar Medium

Solved Problems

6.14. How do astronomers know that stars are formed in giant molecular clouds?

There are three key pieces of evidence that stars are formed in giant molecular clouds. First, these clouds are often the source of significant infrared radiation, implying that there are hot, bright stars shrouded by dust deep within the cloud. Secondly, astronomers often observe young stars very near to giant molecular clouds (i.e., in the center of HII regions). Since the stars are young, they have not yet had time to move far from where they formed, and so probably formed in the molecular cloud. Thirdly, the Hubble Space Telescope pictures of giant molecular clouds yield direct evidence for the presence of protostars, collapsing clouds, and evaporating gaseous globules (EGGs), which are dense star-forming pockets being excavated by brisk winds from hot young stars.

6.15. When is evaporation an important phenomenon in disk evolution? Was this mechanism important in the formation of the solar system?

Evaporation is important only when there is a nearby hot, young star to evaporate the disk material. This was probably not important in the formation of our solar system because the Sun has no nearby neighbors. This is not conclusive, however, since the Sun is older than the entire lifetime of some massive stars.

6.16. What is the significance of the development of a stellar wind?

The onset of the stellar wind effectively stops the infall of more mass onto the protostar. The stellar wind is also responsible for sweeping out the last of the gas remaining in the disk (see Chapters 3 and 7 for more on the topic of stellar winds).

6.17. What happens to turn a protostar into a star?

A protostar is simply a clump of dense hydrogen gas and dust. It does not become a star until it begins to fuse hydrogen in the core, and thus to produce energy.

6.18. Use what you have learned about single star formation to explain how binary stars and clusters of stars might be formed?

These are actually two different processes. Binary stars might be formed from the same clump, in much the same way that Jupiter and the Sun formed from the same cloud. Alternatively, if the cloud had two smaller clumps within it, they might form a binary system.

Clusters of stars, on the other hand, are formed by groups of many clumps in giant molecular clouds, with single or binary stars forming from each clump. All of the clumps are bound together by gravity, however, and so are all the stars that are formed.

6.19. Suppose a clump of a giant molecular cloud rotates once every million years, and has a radius of about 0.05 light years. What is the rotation period of this clump when it has collapsed to the size of the solar system (radius = 40 AU)? (Assume that the mass remains constant.) How does this compare to the orbital periods of Neptune and Pluto? (This method is an estimation, using a simplified model. The situation can be more correctly handled using the moment of inertia of the clump.)

CHAPTER 6   The Interstellar Medium

The speed, v, of a rotating body is given by the rotation speed, ω, times the radius, r,

$$v = \omega r$$

The initial angular velocity is

$$\omega_1 = \frac{2\pi}{T_1} r_1,$$

where T_1 is the period of rotation (10^6 years).

This can be substituted into the angular momentum equation, and set equal to a constant, because angular momentum must be conserved.

$$L = m\omega r^2 = \text{a constant}$$
$$m_1 \omega_1 r_1^2 = m_2 \omega_2 r_2^2$$

but m remains constant,

$$\omega_1 r_1^2 = \omega_2 r_2^2$$
$$\omega_2 = \omega_1 \frac{r_1^2}{r_2^2}$$
$$\omega_2 = \frac{2\pi}{1 \times 10^6 \text{ yr}} \frac{(0.05 \text{ ly})^2}{(40 \text{ AU})^2}$$
$$\omega_2 = \frac{2\pi}{1 \times 10^6 \text{ yr}} \frac{(0.05 \text{ ly} \cdot 63{,}000 \text{ AU/ly})^2}{(40 \text{ AU})^2}$$
$$\omega_2 = \frac{2\pi}{1 \times 10^6 \text{ yr}} \frac{(3{,}163)^2}{(40)^2}$$
$$\omega_2 = \frac{2\pi \cdot 6{,}255}{1 \times 10^6 \text{ yr}} \cong 2\pi / 160 \text{ yr}$$

So the rotation period, finally, is 160 years. This is very close to the orbital period of Neptune (165 years), but much less than the orbital period of Pluto (249 years).

6.20. Why does a collapsing cloud begin to heat up more rapidly as it becomes opaque?

For the gravitational energy released in the collapse to escape the cloud, the photons carrying that energy must be able to escape. As the cloud becomes opaque, the photons become trapped inside the cloud. Thus the energy remains in the cloud and the temperature increases.

6.21. Suppose a cool clump of a giant molecular cloud is exceptionally large, about $r = 10$ pc across, and has a mass, M, of 1×10^3 solar masses. Will this clump collapse to form stars?

For a cloud to collapse, the average velocity must be less than 1/6 the escape velocity. In a cool clump of a giant molecular cloud, most of the particles are hydrogen molecules, and so have a mass of two protons, or 3.4×10^{-24} g. These particles are at a temperature of about 10 K.

$$v_{\text{avg}} = \sqrt{\frac{8kT}{\pi \cdot m}}$$
$$v_{\text{avg}} = \sqrt{\frac{8(1.4 \times 10^{-23} \text{ kg} \cdot \text{m}^2/\text{s}^2/\text{K})(10 \text{ K})}{\pi \cdot (3.4 \times 10^{-27} \text{ kg})}}$$
$$v_{\text{avg}} = 324 \text{ m/s}$$

The escape velocity is given by

$$v_{esc} = \sqrt{\frac{2GM}{r}}$$

$$v_{esc} = \sqrt{\frac{2 \cdot 6.67 \times 10^{-11}\,\text{m}^3/\text{s}^2/\text{kg} \cdot 10^3 \cdot 2 \times 10^{30}\,\text{kg}}{10 \cdot 3 \times 10^{16}\,\text{m}}}$$

$$v_{esc} = 943\,\text{m/s}$$

1/6 of the escape velocity is about 157 km/s. The average velocity is greater than 1/6 the escape velocity, so this clump will not collapse to form stars.

Supplementary Problems

6.22. How many solar mass stars could be formed from a molecular cloud that has a density of 10^{-19} g/cm^3 and a diameter of 0.1 pc? (Assume all the mass is converted to stars.)

Ans. 18

6.23. Suppose a cool clump of a giant molecular cloud has a radius of 1 pc, and a mass of 100 solar masses. Will the clump collapse to form stars?

Ans. Yes

6.24. Suppose a cool clump of a giant molecular cloud has a radius of 10 pc, and a mass of 100 solar masses. Will the clump collapse to form stars?

Ans. No

6.25. The Orion Nebula is about 3′ on the sky. What is its linear diameter?

Ans. 0.4 pc

6.26. The Eagle Nebula contains dust pillars as much as 1/3 pc high. How large is this in miles?

Ans. 6×10^{12} miles, or 6 trillion miles

6.27. In Fig. 6-6, the protostars are not round. Why?

Ans. The accretion disk has begun to form

6.28. Suppose a 10 solar mass protostar developed a stellar wind of 10^{-7} solar masses/year. If this wind continued, how long could the star survive?

Ans. 10^8 years

CHAPTER 6 The Interstellar Medium

6.29. Suppose a clump of a giant molecular cloud rotates once every million years, and has a radius of 0.1 light years. What is the rotation period of this clump when it has collapsed to 40 AU?

Ans. 40 years

CHAPTER 7

Main-Sequence Stars and the Sun

Stars that are fusing hydrogen into helium in their cores are called **main-sequence stars**. Because stars spend most of their lifetimes fusing hydrogen into helium, most stars are main-sequence stars. The Sun, for example, is a main-sequence star. **Red giants** (e.g., Aldebaran, the brightest star in the constellation of Taurus) and **white dwarfs** (e.g., Sirius B, the small companion of the brightest star Sirius) are examples of stars that are not main-sequence stars, and are described in Chapters 8 and 9. The fusion reaction is the primary source of energy in stars and will be discussed in Chapter 8.

Equilibrium of Stars

The lifetime of stars in the main-sequence depend on the properties of the star. The Sun's main-sequence lifetime is estimated at about 10 billion years. Approximately 4.5 billion of those years have already passed.

To remain stable for such a long time, stars must meet two equilibrium conditions.

THERMAL EQUILIBRIUM

The energy produced by the fusion reaction at the core must balance the energy loss through radiation at the surface of the star. Otherwise the star's temperature would not be stable. At the high temperatures of the stellar core, hydrogen atoms are completely stripped of their electrons and the material becomes a **plasma**. For the fusion reaction to occur, the fusing nuclei (which are positively charged) must have enough energy to overcome their mutual electrostatic repulsion. The rate of the fusion reaction increases with increasing temperature and pressure of the core. The hydrogen fusion reaction is discussed in more detail in Chapter 8.

HYDROSTATIC EQUILIBRIUM

The internal pressure of the star must balance the gravitational force. Otherwise the star would either collapse or fall apart. The gravitational force, which tends to collapse the star, is stronger the more massive the star. Therefore, stars with larger mass have higher internal pressures and higher fusion rates and surface temperatures. The **ideal gas law** (see Chapter 1) describes the behavior of main-sequence stars. Collapse of the star decreases the volume and, as a result, the temperature and the gas pressure increase. In addition to the **gas pressure**, the energy created by the fusion reaction at the core is carried outward by photons. The escaping photons collide and push the overlaying material outward, resulting in an outward pressure. This is the **radiation pressure**, which also counteracts the gravitational collapse.

Observable Properties of Stars

THE DISTANCES OF STARS

As the Earth orbits the Sun, nearby stars appear to change their location in the sky relative to the more distant stars (see Fig. 7-1). This is called **parallax**. From the small angle equation (Chapter 1), we have

$$p = 206{,}265 \frac{D}{d}$$

where D is the Earth–Sun distance (1 AU), d is the distance to the star, and p is the parallax angle in arcseconds ($''$). One **parsec** (pc) is 206,265 AU, and the distance in parsecs is given by

$$d(\mathrm{pc}) = \frac{1}{p('')}$$

The largest parallax known is that of Proxima Centauri, which has a parallax of $0.76''$. At present, parallax angles can be measured to an accuracy of $0.001''$. Therefore, the use of the parallax method to determine star distances is reliable for nearby stars only.

Astronomers have found parallax distances to about 400 stars that are accurate to about 1%. Distances to 7,000 stars are known to better than 5% accuracy, and the distances to most stars within 200 pc of the Sun are quite well determined.

CHAPTER 7 Main-Sequence Stars and the Sun

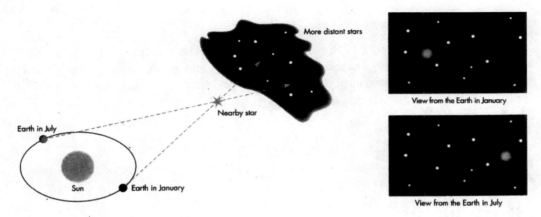

Fig. 7-1. Parallax.

THE BRIGHTNESS OF STARS

Apparent brightness is a term used to indicate how bright a star *appears* to an observer on Earth. The brightness of a star is usually measured (at least in the visible and the infrared) using the **magnitude scale**. The star Vega has magnitude 0, by definition. Fainter stars have larger magnitudes, and brighter stars have smaller magnitudes. Stars fainter than Vega have positive magnitude. Stars brighter than Vega have negative magnitudes. Each increase in magnitude corresponds to a decrease in brightness by a factor of $100^{1/5} \approx 2.5$ (so that an increase of five magnitudes corresponds to a decrease in brightness of 100). The magnitudes can be used to compare the apparent brightnesses of two stars. For two stars of magnitudes m_1 and m_2 and apparent brightnesses b_1 and b_2, respectively, we have

$$\frac{b_1}{b_2} = 2.5^{m_2 - m_1} = 100^{(m_2 - m_1)/5}$$

An intrinsically bright star may appear fainter than an intrinsically dim star if it is much farther away. The intrinsic brightness of a star is called the **luminosity**, L. From the inverse square law (Chapter 1), we have

$$b \propto \frac{L}{d^2}$$

where b is the apparent brightness and d is the distance to the star.

We define absolute magnitude as the apparent magnitude of a star at a distance of 10 pc. Again, an increase of one magnitude corresponds to a decrease in brightness of 2.5. The relationship between apparent magnitude, m, absolute magnitude, M, and distance, d, in parsecs, is given by

$$m = M + 5 \log\left(\frac{d}{10}\right)$$

This equation can be inverted to give the distance in parsecs, provided that both the absolute and the apparent magnitudes are known,

$$d = 10^{(m-M+5)/5}$$

The **luminosity** of a star is the power emitted by the surface of the star at all wavelengths. Luminosity is measured in watts or joules/sec. Assuming that the star emits like a blackbody, we can obtain L as the product of the surface area ($4\pi R^2$) times the power emitted per unit area (using the Stefan-Boltzmann law from Chapter 1):

$$L = 4\pi R^2 \sigma T^4$$

Thus, the luminosity increases with the size and the temperature of the star.

THE MOTION OF STARS

Proper motion refers to the slow motion of stars with respect to other, presumably more distant, stars. The star with the largest proper motion in our sky is Barnard's star, which moves only 10.25" per year (1 degree per 350 years). Most stars move much less than Barnard's star. The velocity of a star has two components: the radial velocity, v_r, and the tangential velocity, v_t.

The radial velocity of a star is measured from the Doppler shift of the emission or absorption lines, as described in Chapter 1. The tangential velocity is obtained by multiplying the proper motion times the distance of the star. The total velocity of the star, with respect to the Sun, is

$$V = \sqrt{v_r^2 + v_t^2}$$

THE TEMPERATURE OF STARS

Studies of the absorption lines in star spectra allowed the grouping of similar stars together. The **spectral classification** led, as our understanding has increased, to a scheme in which stars are ordered by their surface temperature. The spectral classes, from hottest to coolest are: **OBAFGKM**. O stars are the hottest stars and M stars are the coolest stars.

Each spectral class is divided into 10 subclasses, 0–9. An O0 star is the hottest, followed by an O1, O2, etc. O9 is just a little bit hotter than a B0, and so on. Most stars can be classified in this way. Table 7-1 gives spectral characteristics.

The dominant color of the star is related to the temperature of the **photosphere** (the light-emitting layer) by Wien's law (see Chapter 1) and provides a method to determine the star's surface temperature. The spectral features in the classification scheme arise from absorption of the blackbody (continuous emission spectrum) from the star's photosphere by atoms and molecules in the outer layer (the **chromosphere**) of the star. The lines of interest in this classification scheme belong to the visible part of the spectrum (380–750 nm).

A key element in understanding the spectral behavior is that the electrons in the atom can absorb energy and move to a higher energy level. The amount of energy required to move an electron from the ground state to the first excited state is much larger than the energy required to move from the 1st excited level to the 2nd,

CHAPTER 7 Main-Sequence Stars and the Sun

Table 7-1. Characteristics of stars of different spectral classes

Class	Color	Surface temperature (K)	Prominent absorption lines
O	Blue	> 25,000	Ionized helium lines; weak hydrogen Balmer lines; lines of multiply ionized atoms (O, N, C)
B	Blue	11,000–25,000	Neutral helium lines; strong hydrogen Balmer lines
A	Blue–white	7,000–11,000	Very strong hydrogen Balmer lines
F	White	6,000–7,500	Ionized calcium (CaII) lines; strong hydrogen Balmer lines
G	Yellow	5,000–6,000	Strong CaII lines; weak hydrogen Balmer lines; lines of neutral metals
K	Red–orange	3,500–5,000	Weak CaII lines; lines from molecules
M	Red	< 3,500	Lines of neutral elements; strong lines from molecules.

and as a rule the energy steps decrease very rapidly for higher excitations. For example, in the hydrogen atom, the transition from ground state to 1st excited level requires an ultraviolet (UV) photon, from the 1st to 2nd a visible photon, and from 2nd to 3rd an infrared (IR) photon. In our classification scheme, only the 1st to 2nd transition in hydrogen will be relevant. The variation of the prominent absorption lines in the visible part of the spectrum with spectral class, and their relation to the surface temperature of the star is as follows.

Hydrogen Balmer lines. Absorption lines caused by hydrogen atoms that are initially in the first excited state (see Chapter 1) are referred to as Balmer lines. At low surface temperatures, most of the hydrogen atoms are in the ground state, therefore the Balmer lines are weak. As the temperature increases from M through A spectral classes, more and more hydrogen atoms exist in the first excited state, and the strength of the Balmer lines increases. As the temperature further increases, from B to O, more and more hydrogen atoms are excited to the second, third, etc., excited states, and the atoms in the first excited state become fewer. This leads to the progressive weakening of the Balmer lines. Above 12,000 K, most hydrogen atoms become ionized, and have no electron to absorb light.

Helium lines. The energy required to excite a helium atom from the ground state to the first excited state is very large, and requires UV photons, rather than visible photons. At temperatures above 10,000 K, the thermal energy can excite helium atoms to the first excited state. Excited helium atoms can absorb visible photons and move to a higher excitation state. At higher temperatures, more helium atoms exist at higher excitation states, and the energy required for further excitations is low, requiring IR photons. Thus, the neutral helium lines weaken from classes B to O. However, at the high temperature of the O stars, the thermal energy can ionize the helium atom. The remaining electron is more tightly bound to the two positive charges at the nucleus, and the transition to the first excited state requires UV photons. However, at high temperatures the thermal energy is enough to excite the remaining electron to a higher energy level, from which successive transitions to higher levels may be accomplished by absorption of visible photons. O stars often exhibit ionized-helium absorption lines.

Lines from other ions and neutral atoms. In atoms with three or more electrons, each electron experiences attraction by the positive nucleus, and repulsion from the other electrons. When the atom becomes ionized, the remaining electrons are held more tightly, and the energy required for the excitation step is higher, compared with the corresponding steps in the neutral atom. Most elements will readily ionize at the high end of the temperature range of G stars; therefore, absorption by neutral atoms is absent in types G through O. In ions, such as CaII, the transition from the ground state to the first excited state will require UV photons. However, as in the case of helium, thermal energy could excite the outer electron to a higher energy level, from which successive transitions to higher levels may be accomplished by absorption of visible photons. At higher temperatures, the CaII lines, for example, disappear because the thermal collisions excite the electrons to higher energy states, and the amount of energy required for subsequent excitations is small. This happens to CaII as the temperature increases from F through A types. At even higher temperatures, the thermal energy can cause multiple ionizations, and the pattern repeats.

Lines from molecules. These lines are observable in K stars or cooler, because the thermal energy of hotter spectral types is sufficient to overcome the chemical bond that holds the atoms together in the molecule.

THE SIZE (RADIUS) OF STARS

Two properties of a star contribute to its luminosity: the temperature and the size. Hot stars are brighter than cool stars of the same size. Large stars are brighter than small stars of the same temperature. Sometimes a large cool star is brighter than a small hot one. The luminosity class is a way of talking about the radius. There are five luminosity classes, labeled by Roman numerals:

    I. Supergiants
   II. Bright giants
  III. Giants
  IV. Subgiants
   V. Main-sequence dwarfs

The vast majority of stars are dwarf stars. Our Sun is a dwarf star, and if we wanted to describe it in astronomical nomenclature, we would call it a G2V star. Its spectral class is G2, and its luminosity class is V. Be careful not to confuse type V main-sequence stars with white dwarfs, which are much smaller. White dwarfs will be discussed in Chapter 9.

CHAPTER 7 Main-Sequence Stars and the Sun

THE MASS OF STARS

There is a wide variation in the mass of stars. The smallest stars have masses of about 0.08 M_{Sun}. Objects less massive than this never begin hydrogen fusion, and so are never technically considered stars. Objects just below the cutoff often emit energy produced by gravitational collapse, and are called **brown dwarfs**. These objects are brightest in the infrared. The most massive stars known (for example, Eta Carinae) are as large as 150 M_{Sun}.

In general, more massive main-sequence stars are larger (this rule does not necessarily apply to stars that are not main-sequence stars). Also, more massive stars have higher gravity, causing higher temperature and pressure in the core, which speeds the hydrogen fusion reaction. More massive stars produce more energy and are hotter. Since the radius is related to the luminosity, and the mass of the star governs the rate of energy production, there is a mass–luminosity relationship for main-sequence stars:

$$\frac{L}{L_{sun}} = \left(\frac{M}{M_{sun}}\right)^{3.5}$$

where L is the luminosity of the main-sequence star, L_{Sun} is the luminosity of the Sun, M is the mass of the main-sequence star, and M_{Sun} is the mass of the Sun.

THE MAIN-SEQUENCE LIFETIME OF STARS

Once hydrogen burning begins in the core of a protostar, it becomes a main-sequence star. The main-sequence (hydrogen-burning) phase is by far the largest fraction of a star's lifetime, and therefore most stars observed in the sky are main-sequence stars. During this time, stars burn quietly, and change slowly.

The length of time a star remains on the main-sequence depends on the mass. High-mass stars burn hot and die young. This is because when the mass is higher, there is more gravity, and the pressures and temperatures inside the star are higher, which enhances the rate of hydrogen fusion. The relationship between main-sequence lifetime, t, and mass, M, of a star can be expressed mathematically as

$$\frac{t}{t_{Sun}} = \left(\frac{M}{M_{Sun}}\right)^{-2.5}$$

where t_{Sun} is the main-sequence lifetime of the Sun and M_{Sun} is the mass of the Sun. The Sun's main-sequence lifetime is estimated to be 10 billion years. A star 10 times as massive as the Sun has a main-sequence lifetime of only 30 million years. A star 0.1 times as massive as the Sun has a main-sequence lifetime of 3 trillion years.

As stars begin to use up their hydrogen, they increase in size; however, the surface temperature does not change nearly as much. As the radius increases, the luminosity increases as the square of the radius. For example, during the next 5 billion years, the Sun will increase in luminosity by about 60%.

CHAPTER 7 Main-Sequence Stars and the Sun

THE COMPOSITION OF STARS

Observations of spectral lines allow us to determine the composition of stars from the relative strengths of different elemental lines. All stars are primarily composed of hydrogen, with helium the next most abundant element. The abundance of an element in a star generally decreases as you move through the periodic table of the elements.

Solved Problems

7.1. Observations of a star show the hydrogen lines shifted from 486.1×10^{-9} m to 485.7×10^{-9} m. Is the star approaching or receding? How quickly?

Because the stellar lines are shifted to be of shorter wavelength than the reference lines, they are bluer than they should be. This means that the star is approaching. To find out how quickly, use the Doppler equation:

$$\frac{v}{c} = \frac{\Delta \lambda}{\lambda_0}$$
$$v = \frac{0.4 \times 10^{-9}}{486.1 \times 10^{-9}} \cdot 3 \times 10^8 \text{ m/s}$$
$$v = 247{,}000 \text{ m/s}$$
$$v = 247 \text{ km/s}$$

7.2. The luminosity of the Sun is 3.8×10^{-26} watts. How much energy is released after 10 billion years? How much energy is produced in the gravitational contraction of the Sun? Compare these two numbers. Is it reasonable to expect that most of the energy released by the Sun might have come from gravitational contraction?

The luminosity × the time gives the energy released (in joules):

$$E = L \cdot t$$
$$E = (3.8 \times 10^{26} \text{ J/s}) \cdot 1 \times 10^{10} \text{ yr} \cdot 3.16 \times 10^7 \text{ s/yr}$$
$$E = 1.2 \times 10^{44} \text{ J}$$

The amount of energy produced by the collapse of a nebula into a star of mass M and radius R is approximately

CHAPTER 7 Main-Sequence Stars and the Sun

$$E = \frac{GM^2}{R}$$

$$E = \frac{6.67 \times 10^{-11} \frac{m^3}{kg \cdot s^2} \cdot (2 \times 10^{30} \text{ kg})^2}{7 \times 10^8 \text{ m}}$$

$$E = 3.8 \times 10^{41} \text{ J}$$

This is a factor of 1,000 less than the energy released by the Sun over its entire lifetime. Gravitational collapse could not be the main source of the energy released by the Sun.

7.3. Figure 7-2 shows a star field containing Barnard's star, the highest proper motion star known. Indicate which star is Barnard's star.

Fig. 7-2. Two images of Barnard's star, separated by 47 years.

7.4. Parallax is observed as a change in the location of a star relative to the background stars. So is proper motion. Describe how to make an observation of proper motion which includes no parallax, and how to correct for proper motion in a determination of parallactic distance.

To determine the proper motion of a star with no parallactic complications, simply observe the star at the same time of year every year. Since the Earth is always on the same side of the Sun at the same time of year, there is no parallax.
   To correct for proper motion when determining the parallactic distance, first find the proper motion by the method above, then subtract this from the parallax that you have determined.

7.5. Suppose a star has a proper motion of 4"/year, a distance of 2.4 pc, and a radial velocity of 80 km/s. How fast is this star actually traveling through space? Is this fast or slow, compared with the motion of objects on the Earth?

First, convert the proper motion to an angular velocity in "/s.

$$4/\text{yr} \cdot \frac{1\,\text{yr}}{3.16 \times 10^7\,\text{s}} = 1.26^{-7}\,''/\text{s}$$

Now convert this angular velocity to a linear velocity using the small angle equation.

$$v_{\text{pm}} = \frac{\theta \cdot D}{206{,}265}$$

$$v_{\text{pm}} = \frac{1.26 \times 10^{-7} \cdot 2.4 \cdot 3 \times 10^{16}\,''/\text{s} \cdot \text{m}}{206{,}265}$$

$$v_{\text{pm}} = 44{,}000\,\text{m/s} = 44\,\text{km/s}$$

Finally, combine this proper motion in m/s with the radial velocity using the Pythagorean theorem to get the space velocity.

$$V = \sqrt{v_{\text{pm}}^2 + v_{\text{r}}^2}$$

$$V = \sqrt{44^2 + 80^2}$$

$$V = 91\,\text{km/s}$$

This is quite fast compared with most Earth-bound objects.

7.6. Stars move, but the Earth also moves. It rotates on its axis and orbits the Sun. Do these motions contribute to determinations of the radial motion of a star? If so, how might you correct for them?

Yes, these motions do contribute to the determination of the radial motion of stars. One way to correct for this is to always measure the radial motion when the star is close to the meridian. When this is true, all the motion due to the rotation of the Earth is perpendicular to the radial motion of the star. The velocity of the Earth around the Sun is very well known, and easily calculated from the size of the orbit and the length of a year.

7.7. What is the difference between absolute and apparent magnitude?

The absolute magnitude sets all the stars at the same distance and so is a measure of the intrinsic brightness. The apparent magnitude is a measure of how bright it appears to be, given that it may be closer or farther than 10 pc from the Sun.

7.8. How much brighter is a star of first magnitude than a star of fifth magnitude?

$$\frac{b_1}{b_2} = 2.5^{m_2 - m_1}$$

$$\frac{b_1}{b_2} = 2.5^{5-1}$$

$$\frac{b_1}{b_2} = 2.5^4 = 39$$

A first-magnitude star is 39 times brighter than a fifth-magnitude star.

7.9. Given an absolute magnitude of 3.0, find the apparent magnitude at a distance of 5 pc, 10 pc, 15 pc, 20 pc, 50 pc, and 100 pc. Plot these data (Fig. 7-3). How far away must the star be to have an apparent magnitude of 5.0?

CHAPTER 7 Main-Sequence Stars and the Sun

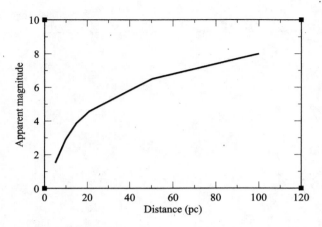

Fig. 7-3. Plot of data below.

$$m = M + 5\log\left(\frac{d(\text{pc})}{10}\right)$$

$$m_{\text{pc}} = 3 + 5\log\left(\frac{5}{10}\right) = 1.5$$

$$m_{10\text{pc}} = 3 + 5\log\left(\frac{10}{10}\right) = 3$$

$$m_{15\text{pc}} = 3 + 5\log\left(\frac{15}{10}\right) = 3.9$$

$$m_{20\text{pc}} = 3 + 5\log\left(\frac{20}{10}\right) = 4.5$$

$$m_{50\text{pc}} = 3 + 5\log\left(\frac{50}{10}\right) = 6.5$$

$$m_{100\text{pc}} = 3 + 5\log\left(\frac{100}{10}\right) = 8$$

From the plot, we can see that a third-magnitude star would need to be at a distance of about 25 pc to have an apparent magnitude of 5.

7.10. Which has the hotter core, a $10\,M_{\text{Sun}}$ star or a $1\,M_{\text{Sun}}$ star? Why?

The $10\,M_{\text{Sun}}$ star has the hotter core. Because there is more mass, the gravitational pressure on the core is higher. This means that the temperature must be higher, which means that hydrogen fuses to helium faster, which means more energy is produced. This increased energy production also contributes to the increase in temperature inside the Sun.

7.11. Place the following spectral types in order of temperature: A, B, F, G, K, M, O

                     O, B, A, F, G, K, M
                     Hot   Cool

7.12. On the chart (Fig. 7-4), label the region where you would expect to find large cool stars, small cool stars, and tiny hot stars.

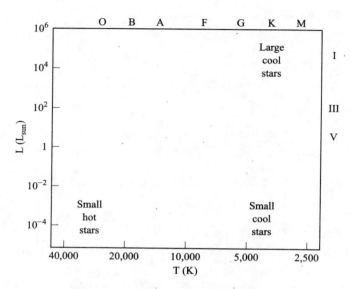

Fig. 7-4. A plot of the relative locations of large cool stars, small cool stars, and small hot stars.

7.13. What is the main-sequence lifetime of a 15 M_{Sun} star?

$$\frac{t}{t_{Sun}} = \left(\frac{M}{M_{Sun}}\right)^{-2.5}$$

$$t = t_{Sun} \cdot \left(\frac{M}{M_{Sun}}\right)^{-2.5}$$

$$t = 1 \times 10^{10} \cdot \left(\frac{15 M_{Sun}}{M_{Sun}}\right)^{-2.5}$$

$$t = 1 \times 10^{10} \cdot (15)^{-2.5}$$

$$t = 1.15 \times 10^{7} \text{ yr}$$

$$t = 11.5 \text{ million years}$$

The lifetime of a star 15 times as massive as the Sun is almost a thousand times shorter.

7.14. Derive an age–luminosity relationship for main-sequence stars.

$$\frac{t}{t_{Sun}} = \left(\frac{M}{M_{Sun}}\right)^{-2.5}$$

$$\frac{L}{L_{Sun}} = \left(\frac{M}{M_{Sun}}\right)^{3.5}$$

Inverting the second equation gives

$$\frac{M}{M_{Sun}} = \left(\frac{L}{L_{Sun}}\right)^{1/3.5}$$

Substituting this back into the first equation gives

$$\frac{t}{t_{Sun}} = \left(\left(\frac{L}{L_{Sun}}\right)^{1/3.5}\right)^{-2.5} = \left(\frac{L}{L_{Sun}}\right)^{-0.7}$$

CHAPTER 7 Main-Sequence Stars and the Sun

For main-sequence stars, the main-sequence lifetime is inversely proportional to the luminosity. This is sensible, because the luminosity indicates how quickly fuel is being burned inside the center of the star. A higher luminosity results in a shorter lifetime.

The Sun

The Sun is a star like many other stars in the sky, but much closer. Because it is so close, we know far more about this star than about any other. Only on the Sun can we see the surface in detail, and begin to understand what processes might be operating in the deepest regions. It is slightly smaller than the average star, but otherwise completely ordinary.

GENERAL STRUCTURE

The general structure of the Sun is shown in Fig. 7-5. In terms of distance from the center, the central 25% of the Sun is actively producing energy by nuclear burning. Outside of this is the radiative zone, which extends to 70% of the Sun. In this radiative zone, energy produced by nuclear fusion is carried outwards by photons. The outer 30% of the Sun is the convective region, where energy is carried outward by hot gas. It takes about 170,000 years for light produced in the center of the Sun to make its way out to the surface. This means that we cannot directly observe what's happening in the center of the Sun now from light.

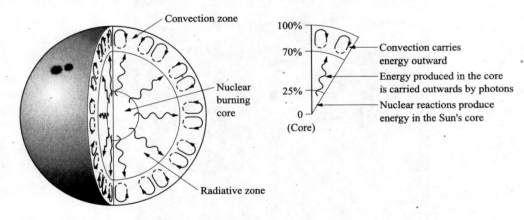

Fig. 7-5. The structure of the Sun.

SOLAR NEUTRINOS

The fusion of hydrogen into helium produces a particle called a **neutrino**. These particles barely interact with ordinary matter, and have a very small mass. They travel outwards through the Sun in about 2 seconds, and in an additional

8 minutes get to the Earth. Observations of solar neutrinos tell us about the current state of the interior of the Sun. Observations of these neutrinos show that there are roughly 1/3 as many of them as theory predicted. This is a very significant question in astrophysics, and the answer may be as simple as a misunderstanding of how many types of neutrinos exist, or as complex as a misunderstanding of how fusion works.

THE PHOTOSPHERE

The **photosphere** is the apparent surface of the Sun. At the bottom of the photosphere, the gas is very dense—so dense that light cannot escape. Outside of this, the density falls off by 50% every 100 km. In a very short distance, the gas is so diffuse that it no longer produces enough light to be seen. All of the light comes from a layer about 200 km thick, which is about 0.03% of the Sun's total radius.

When all the visible light from the photosphere is observed at once, it looks grainy, like the surface of a basketball. This is called **granulation** (Fig. 7-6). This grainy appearance is caused by convection from deeper layers of the Sun. The hot bright material wells up to the surface, and spreads out, then cools and darkens, and falls back to the interior of the Sun. Each of these granules is about 1,000 km across, and lives for only about 15 minutes. Most granules simply fade, but some seem to explode, and push all the other granules about.

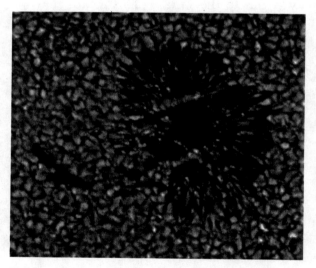

Fig. 7-6. Granules and a sunspot in the photosphere of the Sun. (Courtesy of NASA.)

On a larger scale is a pattern called **supergranulation**. Each supergranule is about 30,000 km across, and contains hundreds of granules. These are the top of convective regions that probably extend all the way to the bottom of the convective zone. These supergranules typically last about 1 day.

CHAPTER 7 Main-Sequence Stars and the Sun

SUNSPOTS

Sunspots are dark marks on the photosphere of the Sun. Often, groups of small spots will merge to form large spots, which can be as large as 50,000 km across (four times the diameter of the Earth), and last for months. The inner, dark part of a sunspot is called the **umbra**. Around this is the **penumbra**. The region surrounding a cluster of sunspots is called an **active region**.

Observations of sunspots can be used to find the rotation period of the Sun. The Sun does not have a single rotation period, but instead rotates at different speeds depending on the solar latitude. Near the equator, the rotation period is about 25 days. Near the poles, it is longer—about 36 days.

Sunspots are dark because the umbra is about 2,000 K cooler than the rest of the photosphere. Sunspots are not actually dim, however. A sunspot against the night sky would appear about 10 times as bright as the full moon. Sunspots only seem dim because the rest of the Sun is so bright.

Sunspots are regions of strong magnetic fields. These magnetic fields trap the particles that have bubbled up to the surface by convection. These particles can't cross the magnetic field lines, and so they can't move out of the way to allow new hot material to bubble up. The particles cool off, and therefore darken, but remain in place.

The magnetic field comes out of the Sun at one location, and goes back into the Sun at another. This means that sunspots always come in pairs. One spot will have north polarity like the north pole of a bar magnet, and the other will have south polarity.

THE SUNSPOT CYCLE

Daily records of the number of sunspots on the Sun have been kept since about 1850. Other records exist (not daily ones) back to about 1610. Figure 7-7 shows the number of sunspots over time.

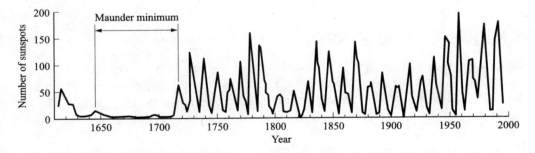

Fig. 7-7. The solar cycle. (Plot courtesy of NASA/YOHKOH.)

The number of sunspots rises and falls quite regularly, with a cycle of about 11 years. The low numbers of sunspots found around 1800 and 1900 imply that there might be another cycle about 100 years long.

Particularly interesting is a lack of sunspots between 1640 and 1700. This is called the Maunder Minimum, and coincided with a time of very cold climate in North America and Western Europe, known as the Little Ice Age. It is not clear that the two things are related, but it is a possibility.

Not only the number but also the location of sunspots varies over the cycle. Sunspots appear about 30 N or S of the solar equator at the beginning of the cycle, and gradually migrate toward the equator near the end of the cycle. This pattern yields the famous "butterfly diagram," shown in Fig. 7-8.

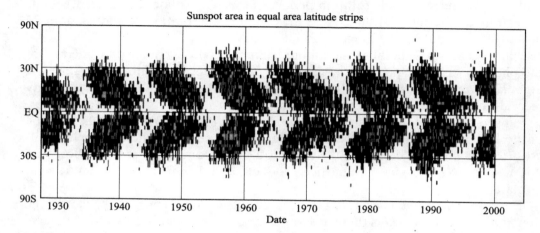

Fig. 7-8. The butterfly diagram. At the beginning of each cycle, the spots are located far from the solar equator. As the cycle progresses, the spots form closer to the equator. (Courtesy of NASA/YOHKOH.)

The differential rotation of the Sun causes the magnetic field lines to become twisted tightly around the Sun. Eventually, the lines become so tightly twisted that they poke through the surface. The pairs of sunspots migrate—each sunspot in opposite directions—and eventually flip the magnetic field of the Sun. This causes the polarity of the sunspots to reverse in each cycle.

PROMINENCES AND FLARES

During most total solar eclipses, bright clouds of gas can be seen rising 50,000 km or more from the photosphere (Fig. 7-9). These are called prominences, and sometimes last for 2–3 months. When prominences are seen against the surface of the Sun, they are called filaments, because they appear in projection as long thin strands.

Prominences are cooler than the photosphere. Usually, prominences rise up through the outermost layers of the Sun until they burst open, and fling mass out into space. These blasts can be traveling as fast as 1,000 km/s. This is called a coronal mass ejection, and there are one or two of these each day on average.

Solar flares are much more energetic and massive than prominences. These take place in active regions when the magnetic field structure changes abruptly. Large numbers of ions and electrons go flying through the corona, heating the gas, and

CHAPTER 7 Main-Sequence Stars and the Sun

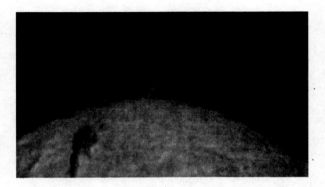

Fig. 7-9. Solar prominence on the limb of the Sun. (Courtesy of NASA.)

causing it to radiate brightly, mostly in X-rays and ultraviolet light. When this radiation intersects the Earth, it interacts with our ionosphere, disrupting long-range radio communication.

THE CHROMOSPHERE

Just above the photosphere lies the chromosphere, a very thin atmosphere around the Sun. Only about $10^{-12}\%$ of the Sun's mass is in the chromosphere, and so it is transparent to all wavelengths. A very few strong absorption lines can be observed, such as H-alpha. The chromosphere is ragged in structure. This is caused by spicules, or hot jets of gas that shoot upward from the surface at about 50,000 miles per hour. Each of these lasts about 5 minutes. These cover a few percent of the Sun's surface, as shown in Fig. 7-10, in which the dark, linear structures are the spicules.

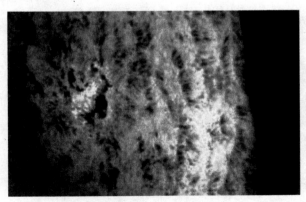

Fig. 7-10. Spicules in the chromosphere. (Courtesy of NASA.)

Spicules always lie along the magnetic field lines of the Sun, and can be a good indicator of the direction of the magnetic field in a given location on the Sun. Other than spicules, the chromosphere is pretty much empty.

THE CORONA

Above the chromosphere is the corona (Fig. 7-11), a blue haze easily observed around the disk of the Moon during a solar eclipse. Much of the light in the corona was produced by the photosphere, and then scattered towards us by the electrons in the corona. Some of it comes from highly ionized atomic transitions. The temperature of the corona is very high, more than 1 million K. The transition between the cool chromosphere (5,000 K) and the superheated corona is abrupt; only a few thousand kilometers separate the two regions.

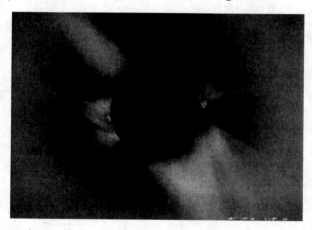

Fig. 7-11. An optical image of the solar corona. An opaque disk was used to block the light of the photosphere, so that the corona would be visible. (Courtesy of NASA.)

One explanation for the high temperature of the corona is that the magnetic field interactions heat the particles. Another possibility is that only the electrons that have enough energy to get out of the chromosphere produce the corona.

In an X-ray picture of the corona (Fig. 7-12), locations where there is almost no corona at all are shown: these are coronal holes. The bright parts are places where

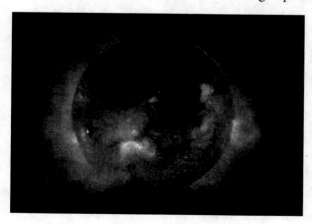

Fig. 7-12. An X-ray image of the solar corona, showing coronal holes. (Courtesy of NASA.)

CHAPTER 7 Main-Sequence Stars and the Sun

the magnetic field has trapped hot material, which occurs mostly over active regions.

THE SOLAR WIND

The corona is constantly losing gas into the solar system, and this material is called the solar wind. It flows away from the Sun at about 450 km/s, and takes 4 days to reach the Earth. The Voyager and Pioneer spacecraft have shown that the solar wind continues to move outward to more than 50 AU from the Sun. The edge of the solar wind is estimated to be at about 100 AU. This "boundary" is called the heliopause and indicates the end of the Sun's direct influence on its environment. Outside of this boundary, the only effect the Sun can have is via the light that it produces.

HELIOSEISMOLOGY

Helioseismology is the study of the oscillations of the Sun through Doppler shifts in the photosphere. This is just like studying earthquakes on the Earth (see Chapter 3). Observations of seismic waves on Earth have told us about the Earth's interior, and observations of seismic waves on the Sun are telling us about the Sun's interior. As on the Earth, short-wavelength vibrations sample the surface, and long-wavelength vibrations sample the depths. Carefully analyzing the periods of these vibrations reveals the temperature and composition of the inner layers of the Sun.

Solved Problems

7.15. The total luminosity of the Sun is 4×10^{26} watts. Calculate the total number of watts that fall on a 1 m^2 sheet of paper on the Earth's surface. (Recall the inverse square law.) Compare this number to the power output of a standard light bulb.

The total luminosity of the Sun is the amount of light emitted by the entire surface area of the Sun. This means that the flux here at Earth is the total power of the Sun, divided by the surface area of a sphere the size of the Earth's orbit:

$$f = \frac{3.80 \times 10^{26} \text{ W}}{4\pi (1 \text{ AU})^2}$$

$$f = \frac{3.80 \times 10^{26} \text{ W}}{4\pi (1.50 \times 10^{11} \text{ m})^2}$$

$$f = 1{,}350 \text{ W/m}^2$$

The total amount of power incident on a 1 m² piece of paper is 1,350 W. This calculation ignores atmospheric opacity, which removes most of the high-energy flux, and much of the infrared flux. The energy flux is equivalent to the energy detected about 7 cm from the filament of a 100 W light bulb.

7.16. Examine Fig. 7-8. What is the period of sunspot activity (how long between maxima)? When is the next solar maximum? When is the next solar minimum?

The period of sunspot activity is 11 years. The last solar maximum was in the year 2000, and so the next solar maximum will be in 2011. The next solar minimum is in 2005.

7.17. Why does the Sun appear to have a sharp edge?

The Sun appears to have a sharp edge because it goes from being opaque to being completely transparent in only a few hundred kilometers. This is such a small fraction of the size of the Sun that it appears to be instantaneous.

7.18. What causes sunspots?

Magnetic field lines trap material which has been convected to the surface. This trapped material cools, and becomes darker than the surrounding material, causing a "spot," that persists for many days.

7.19. How hot (approximately) is the penumbra of a sunspot?

The umbra of a sunspot is about 3,000 K. The photosphere is about 5,000 K. The penumbra is about halfway between the two in brightness, and so is probably about 4,000 K.

7.20. The mass loss rate of the Sun is about $3 \times 10^{-14} \, M_{\text{Sun}}/\text{yr}$. How much mass is intercepted by the Earth each day? (For simplicity, assume the mass loss is spherical.)

To solve this problem we must know what fraction of the solar wind is intercepted by the Earth's area. In other words, how big is the disk of the Earth compared with a sphere of radius 1 AU?

$$\frac{A_{\text{Earth}}}{A_{\text{sphere}}} = \frac{\pi \cdot R_{\text{Earth}}^2}{4\pi \cdot R_{\text{sphere}}^2}$$

$$\frac{A_{\text{Earth}}}{A_{\text{sphere}}} = \frac{(6 \times 10^6 \, \text{m})^2}{4(1.5 \times 10^{11} \, \text{m})^2}$$

$$\frac{A_{\text{Earth}}}{A_{\text{sphere}}} = 4 \times 10^{-10}$$

This is the fraction of the wind that will be intercepted by the Earth.

$$M = 3 \times 10^{-14} \, \frac{M_{\text{Sun}}}{\text{yr}} \cdot 4 \times 10^{-10}$$

$$M = 1.2 \times 10^{-23} \, \frac{M_{\text{Sun}}}{\text{yr}}$$

Now, convert years to days:

$$M = 4.4 \times 10^{-21} \, \frac{M_{\text{Sun}}}{\text{day}}$$

CHAPTER 7 Main-Sequence Stars and the Sun

Finally, convert to kg from solar masses:

$$M = 8.8 \times 10^9 \, \frac{\text{kg}}{\text{day}}$$

Nearly 9 billion kilograms of mass are intercepted by the Earth each day!

7.21. Calculate the percent change in your weight each day due to the increase of the Earth's mass because of the solar wind.

The force of gravity (weight) is given by

$$F = \frac{GmM}{R^2}$$

In this problem, all of the terms on the right-hand side remain constant except M, the mass of the Earth.

$$\frac{F_{day2}}{F_{day1}} = \frac{\frac{GmM_{day2}}{R^2}}{\frac{GmM_{day1}}{R^2}}$$

$$\frac{F_{day2}}{F_{day1}} = \frac{M_{day2}}{M_{day1}}$$

$$\frac{F_{day2}}{F_{day1}} = \frac{M_{day1} + 8.8 \times 10^9 \text{ kg}}{M_{day1}}$$

$$\frac{F_{day2}}{F_{day1}} = \frac{M_{day1}}{M_{day1}} + \frac{8.8 \times 10^9 \text{ kg}}{M_{day1}}$$

$$\frac{F_{day2}}{F_{day1}} = 1 + \frac{8.8 \times 10^9}{6 \times 10^{24}}$$

$$\frac{F_{day2}}{F_{day1}} = 1 + 1.5 \times 10^{-15}$$

Your weight is increased by 0.000000000000015% each day due to the increase of the mass of the Earth from intercepted particles in the solar wind. The 8.8 billion kilograms of material landing on the Earth each day has very little effect on anything at all.

7.22. What is the difference between a sunspot and an active region?

A sunspot is a particularly cool location on the Sun, where the magnetic field traps upwelling material. An active region is the area surrounding a collection of sunspots, where the magnetic field is strong. One way to say this is that the sunspots are the symptoms of the active region.

7.23. Figure 7-13 shows two images of the Sun, taken x days apart. What is the rotation period of the Sun (at the equator)?

Using a ruler, find the diameter of the Sun and the horizontal distance from the limb of the central bright spot in each image:

$$D_{Sun} = 4.5 \text{ cm}$$
$$X_{day1} = 2 \text{ cm}$$
$$X_{day2} = 2.5 \text{ cm}$$

CHAPTER 7 Main-Sequence Stars and the Sun

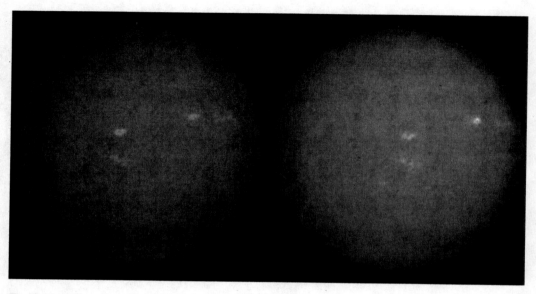

Fig. 7-13. Two pictures of the Sun, taken 1 day apart.

From the diameter, we can calculate the circumference of the Sun in these images:

$$C = 2\pi R$$
$$C = \pi D$$
$$C_{Sun} = 14.14 \, cm$$

Dividing the circumference by the speed of the spot (number of cm/day the spot has moved) gives the rotation period in days:

$$P = \frac{14.14 \, cm}{0.5 \frac{cm}{day}}$$
$$P = 28 \, days$$

This is in fairly good agreement with the accepted equatorial rotation period of the Sun (25 days). The measurement technique is crude, and uses small images, which can't be measured better than about 0.1 cm, since the spots are extended, and the limb of the Sun is not sharp. If we had measured a change in location of 0.6 cm/day, the rotation period of the Sun would be 23.6 days, implying an error in our measurements of about 16%. This is "good agreement" in most astronomical situations.

7.24. What is the absolute magnitude of the Sun? (The apparent magnitude is -26.5.)

$$m = M + 5\log\left(\frac{d(pc)}{10 \, pc}\right)$$
$$M = m - 5\log\left(\frac{d(pc)}{10 \, pc}\right)$$
$$M = -26.5 - 5\log\left(\frac{1/206{,}265}{10}\right)$$
$$M = 5.1$$

The absolute magnitude of the Sun is 5.1. This is brighter than the naked-eye limit. If we were 10 pc from the Sun, it would appear as another unremarkable point of light in the sky.

CHAPTER 7 Main-Sequence Stars and the Sun

7.25. If you were to move to Europa, how large would the Sun be in your sky?

The distance from Jupiter (and therefore Europa, approximately!) to the Sun is 5.2 AU. The diameter of the Sun is 1.4×10^9 m. Using the small angle formula (see Chapter 1),

$$\frac{d}{D} = \frac{\theta}{206{,}265}$$

$$\theta = 206{,}265 \cdot \frac{d}{D}$$

$$\theta = 206{,}265 \cdot \frac{1.4 \times 10^9}{5.2 \cdot 1.5 \times 10^{11}}$$

$$\theta = 370''$$

$$\theta = 0.1°$$

This is about 1/5 the size of the Sun as viewed from Earth.

7.26. Assume that the Sun has been producing the same wind since it began burning hydrogen (about 4.5 billion years ago). How far has the initial wind traveled? When did the solar wind reach Alpha Centauri? (Assume the wind remains at the same velocity always!)

The solar wind travels at about 450 km/s; so in 4.5 billion years it has traveled

$$d = v \cdot t$$

$$d = 450 \text{ km/s} \cdot 4.5 \times 10^9 \text{ yr} \cdot 3.16 \times 10^7 \text{ s/yr}$$

$$d = 6.4 \times 10^{19} \text{ km}$$

$$d = 4.3 \times 10^8 \text{ AU}$$

If we could travel amongst the stars, we would be able to find particles from the solar wind 430 million AU away.

Alpha Centauri is located 1.33 pc away, or 3.9×10^{13} km. At 450 km/s, it would take the solar wind

$$t = d/v$$

$$t = \frac{3.9 \times 10^{13} \text{ km}}{450 \text{ km/s}}$$

$$t = 8.7 \times 10^{10} \text{ s}$$

$$t = 2{,}700 \text{ yr}$$

to reach Alpha Centauri. Particles from the Sun may have reached Alpha Centauri fewer than 3,000 years after the wind began.

7.27. Determine the kinetic energy (KE) of a single proton in the solar wind. How does this compare with the energy of an X-ray or gamma-ray?

The kinetic energy of a particle is given by

$$\text{KE} = \frac{1}{2}mv^2$$

$$\text{KE} = \frac{1}{2} \cdot 1.7 \times 10^{-27} \text{ kg} \cdot (450{,}000 \text{ m/s})^2$$

$$\text{KE} = 1.7 \times 10^{-16} \text{ J}$$

CHAPTER 7 Main-Sequence Stars and the Sun

An X-ray has a frequency of about 10^{18} Hz. The energy of a photon is given by

$$E = h \cdot f$$
$$E = (6.626 \times 10^{-34} \text{ W} \cdot \text{s}^2) \cdot 10^{18} \text{ Hz}$$
$$E = 6.626 \times 10^{-16} \text{ J}$$

This is only a factor of about three larger than the kinetic energy of the proton! Protons in the solar wind have enough energy to disrupt cells and cause mutations. Fortunately, the Earth is surrounded by a magnetic field that helps to protect us from these high-velocity charged particles.

7.28. How many people are required to fill the volume of the Earth? How many Earths are required to fill Jupiter? How many Jupiters are required to fill the Sun? How many Suns are required to fill the solar system? (Assume all of these solar system objects are spherical.)

First, estimate the volume of a person. Assume a person is 1.5 m tall, 0.3 m wide, and 0.2 m thick. Therefore the volume of a typical person is about 0.1 m³. The volumes of the solar system objects are

$$V = \frac{4}{3}\pi \cdot R^3$$
$$V_{\text{Earth}} = \frac{4}{3}\pi \cdot (6.4 \times 10^6 \text{ m})^3 = 1 \times 10^{21} \text{ m}^3$$
$$V_{\text{Jupiter}} = \frac{4}{3}\pi \cdot (72 \times 10^6 \text{ m})^3 = 1.6 \times 10^{24} \text{ m}^3$$
$$V_{\text{Sun}} = \frac{4}{3}\pi \cdot (7 \times 10^8 \text{ m})^3 = 1.4 \times 10^{27} \text{ m}^3$$
$$V_{\text{solar system}} = \frac{4}{3}\pi \cdot (40 \text{ AU} \cdot 1.5 \times 10^{11} \text{ m/AU})^3 = 9 \times 10^{38} \text{ m}^3$$

Therefore, it takes about 10^{22} people to fill the Earth, 1,600 Earths to fill Jupiter, about 1,000 Jupiters to fill the Sun, and 9×10^{11} Suns to fill the solar system. The point here is that people are very small, and the Sun and solar system are very big. Also, the solar system is mostly empty! Less than 1 part in 10^{11} (100 billion) is actually filled with matter.

7.29. How large is a solar prominence compared with the size of the Earth and with the distance from the Earth to the Moon?

A typical solar prominence is about 50,000 km high. The Earth is about 12,700 km across. Approximately four Earths would fit within the height of a solar prominence. The distance from the Earth to the Moon is about 384,000 km. A solar prominence would reach nearly one-eighth of the way to the Moon.

7.30. Are prominences bright or dark when seen against the disk of the Sun? Why?

Prominences are dark when seen against the disk of the Sun, because they are cooler than the photosphere, and cooler things observed against a hot background look dark. Prominences are not actually dark, however, and when seen against empty space (during an eclipse, for example), they appear quite bright.

7.31. If you were traveling through space, and entered the heliopause of a Sun-like star, how would you know?

CHAPTER 7 Main-Sequence Stars and the Sun

The heliopause marks the location where the solar wind becomes so diffuse that it is indistinguishable from the interstellar medium. If you were approaching a star, you would recognize the heliopause because the density of the medium you were traveling through would begin to rise steadily.

Supplementary Problems

7.32. Main-sequence star X is 10 times as luminous as main-sequence star Y. How do their main-sequence lifetimes compare?

Ans. Star X lives 20% as long as star Y

7.33. Main-sequence star A is 10 times as massive as main-sequence star B. How do their main-sequence lifetimes compare?

Ans. Star A lives 3% as long as star B

7.34. Which is more important to the lifetime of the star, the luminosity or the mass? Explain why.

Ans. Mass, because it determines the luminosity

7.35. How many Moon diameters ($d = 3{,}476$ km) would it take to make the diameter of a large sunspot?

Ans. 14.4

7.36. Table 7-2 gives data for the 15 nearest stars. Study the table and answer the following questions.
(a) Which star (after the Sun) is brightest?
(b) Which star is faintest?
(c) Which star has the highest luminosity?
(d) Which star has the lowest luminosity?
(e) Which star is most like the Sun?
(f) Why does Sirius B not have a spectral class listed?
(g) How many stars are M type?
(h) How many stars are G type?
(i) How many stars are another type?
(j) Using just the information in this table, would you say the Sun is a "typical" star?

Ans. (a) Sirius A; (b) Wolf 359; (c) Sirius A; (d) Wolf 359; (e) Alpha Centauri A; (f) it is not a main-sequence star; (g) 9; (h) 2; (i) 4; (j) no, it is brighter and yellower than the average star

7.37. Table 7-2 gives data for the 15 nearest stars. What is the parallactic angle of Wolf 359, Sirius, and Proxima Centauri?

Ans. 0.42″, 0.38″, 0.77″

Table 7-2. Data for the 15 nearest stars

Name	Distance (pc)	Spectral type	Apparent magnitude	Absolute magnitude
Sun	—	G2	−26.7	4.8
Proxima Centauri	1.30	M5	11.1	15.5
Alpha Centauri A	1.33	G2	0.0	4.4
Alpha Centauri B	1.33	K0	1.3	5.7
Barnard's Star	1.83	M4	9.6	13.2
Wolf 359	2.39	M6	13.4	16.6
BD +36 2147	2.52	M2	7.5	10.5
L276-8A	2.63	M6	12.4	15.3
L276-8B	2.63	M6	13.2	16.1
Sirius A	2.63	A1	−1.4	1.5
Sirius B	2.63	White dwarf	8.4	11.3
Ross 154	2.93	M3	10.5	13.1
Ross 248	3.17	M5	12.3	14.8
Eta Eri	3.27	K2	3.7	6.2
Ross 128	3.32	M4	11.1	13.5

7.38. Two stars differ in magnitude by three magnitudes. What is the ratio of their brightnesses?

Ans. 15.6

7.39. You observe a star with an apparent magnitude of 4. Its parallactic angle is 0.32″. How far away is the star?

Ans. 3.1 pc

7.40. What is the absolute magnitude of the star in Problem 7.38?

Ans. 6.5

7.41. What is the luminosity of a main-sequence star of 8 M_{Sun}?

Ans. 1,400 L_{Sun}

7.42. What is the mass of a star with luminosity 8 L_{Sun}?

Ans. 1.8 M_{Sun}

CHAPTER 7 Main-Sequence Stars and the Sun

7.43. What is the main-sequence lifetime of a star of 4 M_{Sun}?

Ans. 300 million years

7.44. A message is received from aliens saying their star is about to leave the main-sequence after 8 billion years. What was the mass of their star?

Ans. 1.1 M_{Sun}

7.45. How many times do the polar regions on the Sun rotate before they "lap" the equator?

Ans. 0.7 times

7.46. What is the difference between a supergranule and a large sunspot?

Ans. Supergranules are locations where material bubbles to the surface. Sunspots are regions where the magnetic field bubbles to the surface.

7.47. How many Earth diameters make up the height of a solar prominence?

Ans. Nearly 4

7.48. Use Figs 7-8 and 7-9 to determine whether sunspots appear farthest from the equator during solar maximum, solar minimum, or in between.

Ans. Solar maximum

CHAPTER 8

Stellar Evolution

Stars evolve. They are born from big clouds of dust and gas, and live most of their lives quietly burning hydrogen on the main sequence. Then they shed mass back into the interstellar medium, either gently or explosively, depending on their main-sequence mass. A small cinder is left over, which cools and contracts, sometimes to extremely high densities.

Why Do Stars Evolve?

Stars produce energy by nuclear fusion, which converts lighter elements to heavier elements (e.g., hydrogen to helium). This changes the chemical composition in the core of the star, which changes the energy production rate, the energy escape rate, and the internal pressure. All of these changes make the star look different from the outside. We will now look in more detail at each of these changes.

ENERGY PRODUCTION

There are three kinds of energy production that can take place in the core of a star: hydrogen fusion, fusion of heavier elements, and gravitational collapse.

HYDROGEN FUSION

Hydrogen fusion occurs at temperatures of ~10 million degrees kelvin (K). At these temperatures, the protons that are the nuclei of hydrogen atoms have enough energy to overcome the mutual repulsion due to charge. (Recall that two particles with the same charge repel each other, but particles with opposite charges attract each other.) These protons can enter the **proton–proton (p–p) chain** (Fig. 8-1).

CHAPTER 8 Stellar Evolution

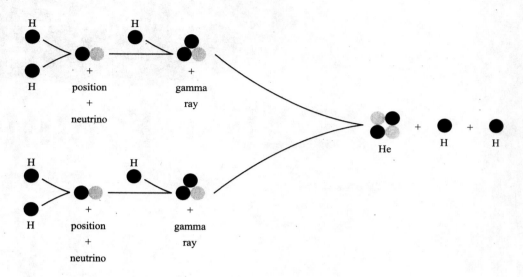

Fig. 8-1. The proton–proton chain.

In Fig. 8-1 you can see the production of **positrons** (a form of antimatter). Positrons are exactly like electrons except that they have a positive charge. A positron and an electron can annihilate, and produce energy in the form of gamma rays.

Neutrinos are particles that have no charge, travel very near the speed of light, and have a very small amount of mass when at rest. Neutrinos barely interact with ordinary matter at all. There are 10 million neutrinos passing through each centimeter of your skin each second. Most neutrinos pass through the entire Earth without interacting at all. We observe these neutrinos with special detectors buried deep in the Earth, which are essentially big chambers filled with a chlorine-containing fluid. Occasionally (about twice per day), a neutrino will interact with a chlorine atom, and turn it into radioactive argon. These argon atoms are collected and counted to estimate the number of neutrinos passing through the fluid.

Typical protons in the center of the Sun travel at 1 million km/h at temperatures of 10 million K. For two protons to interact, and begin the p–p chain, they must pass within 10^{-15} m of one another. Therefore, the probability of interaction is very low. If you observe an individual proton, it will take 5 billion years for that proton to become involved in the p–p chain. However, there are so many protons in the Sun that even though it takes 5 billion years for an individual proton to interact, there are still 10^{34} protons interacting every second.

EINSTEIN'S MASS–ENERGY RELATION

The energy released from fusion comes from the transformation of mass into energy. The six hydrogen nuclei (protons) entering the proton–proton chain have more mass than the helium nucleus and two hydrogen nuclei that leave the p–p chain. The mass of the interacting protons is

CHAPTER 8 Stellar Evolution

$$6m_p = 6 \cdot (1.674 \times 10^{-27})\,\text{kg} = 10.044 \times 10^{-27}\,\text{kg}$$

The products (2 protons and 1 helium) have mass equal to

$$2m_p + m_{He} = 2(1.674 \times 10^{-27}) + 6.643 \times 10^{-27}\,\text{kg} = 9.991 \times 10^{-27}\,\text{kg}$$

The mass difference of 0.053×10^{-27} kg is converted to energy according to Einstein's relation $E = mc^2$. This is a very small amount of energy. But, if we multiply by 10^{34} per second, we get a large energy output, equal to the energy output of the Sun. The mass of the positron is very small and was ignored in the calculation.

In stars more massive than the Sun, another process operates which uses carbon as a catalyst to assist the fusion of hydrogen into helium. This process is called the CNO cycle, because carbon is turned into nitrogen, then into oxygen, then back into carbon plus helium. The carbon repeats the cycle. The end product of the cycle is the transformation of four hydrogen nuclei into one helium nucleus.

FUSION OF HEAVIER ELEMENTS

Other elements have more protons than hydrogen, and so are even more positively charged, so that even higher temperatures and pressures are required to cause fusion. For example:

1. *Helium burning.* Helium is fused into carbon at 100 million K.
2. *Carbon burning.* Carbon begins to be burned at 500 million to 1 billion K.
3. *Iron burning.* Elements whose mass is greater than iron, or iron itself, all use energy when they burn. In other words, more energy goes in than comes out. This means that iron fusion or fusion of heavier elements cannot make a star shine. These heavy elements are produced when the star is no longer steadily producing energy and explodes (see Supernovae, below).

CONTRACTION AND COLLAPSE

Gravitational potential energy is produced when mass falls onto the star. The total amount of gravitational energy in a star is given by

$$E_{tot} = \frac{GM^2}{R}$$

where G is the gravitational constant (equal to 6.67×10^{-11} Nm²/kg), M is the mass of the star, and R is its radius. For example, if the Sun were to shrink to the size of the Earth, 3 billion years worth of solar energy would be released. This process of contraction is usually slow, because energy production in the inner regions balances gravity pulling inward.

ENERGY ESCAPE RATE

The energy escape rate is governed by the **opacity** ("opaque-ness"), which changes with the chemical composition. As the star's opacity increases, the photons produced cannot escape the star. This raises the temperature, and therefore the pressure. Usually, a star's outer layers will swell in response to this increased pressure.

IDEAL GAS PRESSURE

The interior of main-sequence stars can be described by the ideal gas law (see Chapter 1). As a star evolves, the temperature and pressure must always be in equilibrium. If the pressure goes up, the temperature must go up and, conversely, if the temperature goes down, the pressure also falls. As an example, consider what happens when a star runs out of hydrogen in the core. With no fuel to burn, the temperature falls, so the pressure falls, and the core of the star contracts under the influence of gravity. As the core contracts, the pressure increases, and therefore the temperature rises. If the mass of the star is high, the temperature and pressure increases may initiate fusion of helium. When helium ignites, the temperature rises some more, and so does the pressure, and the star's outer layers expand. Changing the internal pressure by changing the chemical composition has a large effect on the way the star appears on the outside.

DEGENERATE GAS PRESSURE

When the core of a star becomes very dense, it can no longer be described by the ideal gas law. The temperature of such star can increase without increasing the pressure. This is referred to as "degenerate gas." Degenerate gas is discussed further in Chapter 9.

How Do Stars Evolve?

THE H-R DIAGRAM

The Hertzsprung-Russell diagram (**H-R diagram**) is a plot of temperature versus luminosity for many different stars. An example is shown in Fig. 8-2. Hot stars are on the left, and low-luminosity stars are at the bottom. So white dwarfs, which are hot and small, are in the lower left, and red giants, which are cool and large, are in the upper right.

When stars are plotted in this way, patterns begin to emerge. Most stars fall in a band stretching from the upper left to the lower right. These are the main-sequence stars discussed in Chapter 7. Because most observable stars fall in this band, we know that this is the longest portion of a star's lifetime.

CHAPTER 8 Stellar Evolution

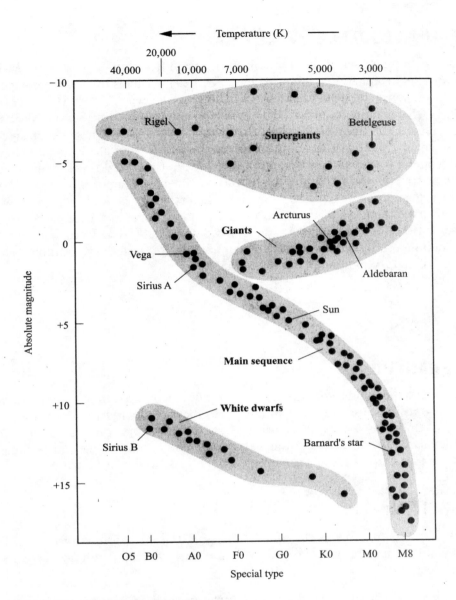

Fig. 8-2. An example H-R diagram.

MAIN-SEQUENCE EVOLUTION

During their main-sequence lifetime, stars are in hydrostatic and thermal equilibrium (see Chapter 7) and therefore do not change appreciably. The star burns hydrogen in the core, and as the hydrogen burns, it leaves a pile of helium ash in the center. The star may change slightly in luminosity (by a factor of two or so), or in size (by less than a factor of two), and gives the energy up to the rest of the Universe mainly in the form of photons and neutrinos.

POST-MAIN-SEQUENCE EVOLUTION

A star leaves the main sequence, by definition, when it exhausts the hydrogen in the core, and begins burning helium instead. As the star leaves the main sequence, it begins to change in spectacular ways. There are two ways for stars to evolve after leaving the main sequence. The process is determined by the star's mass. Stars with mass less than 8 M_{Sun} lose their outer layers slowly. These outer layers are illuminated by the hot core, and become ionized, glowing shells. These shells are called **planetary nebulae**, not because they have anything to do with planets but because they are round fuzzy disks when viewed through a small telescope. Stars with mass larger than 8 M_{Sun} lose their outer layers explosively, in a supernova. These outer layers are also illuminated by the exposed core, and become ionized, but the velocities of these **supernova remnants** is much different from that of the planetary nebulae. In general, supernova remnants have quite large outflow velocities, while planetary nebulae are less energetic.

Stars < $8M_{Sun}$

GIANT BRANCH

When stars leave the main sequence, they begin burning hydrogen in an expanding thin shell around the core. Helium ash falls onto the core, increasing the mass and the density of this core. The hydrogen-burning shell grows in radius as the interior fuel is depleted. The outer layers of the star swell and cool, and so the star becomes a **red giant**, and occupies the giant branch on the H-R diagram (see Fig. 8-2).

HELIUM FLASH

If a star is about as massive as the Sun (< 2 M_{Sun}), then it will undergo a helium flash while on the giant branch. During a helium flash, the star undergoes a series of changes deep within:

1. As helium ash from the hydrogen-burning shell falls onto the core, the helium core gains mass, and shrinks, increasing in density.
2. The electrons in the core become degenerate and the temperature increases without increasing the pressure, so the core can no longer expand.
3. The hydrogen-burning shell makes more ash; the helium core continues to grow.
4. The temperature in the degenerate core rises to 100 million K; the mass in the degenerate core accumulates to 0.6 M_{Sun}.
5. Helium fusion begins suddenly in the core.
6. Because the core is made of degenerate electrons, the pressure and temperature are not related. Even though the core gets hotter, the pressure stays the same, and the core does not expand.

CHAPTER 8 Stellar Evolution

7. However, the higher temperature increases the rate of helium burning, further increasing the temperature.
8. Temperature and burn rate increase until $T \sim 300$ million K. For a few minutes, the energy production of the core of the star is about 100 times the entire energy production of the Milky Way Galaxy.
9. At 300 million K, the electrons stop being degenerate, the interior expands, and the temperature and density drop.
10. This is all deep inside. Most of this energy is consumed in puffing up the core. The outside of the star contracts somewhat, raising the temperature, so that it becomes a yellow giant.

Stars with mass greater than $2\,M_{Sun}$ do not have helium flashes. The temperature rises too fast for the electrons to become degenerate. They **do** begin to burn helium in their cores, just not explosively. Once helium burning begins in any star, there are two sources of energy—the helium-burning core, which is turning helium into carbon and a little bit of carbon into oxygen, and the hydrogen-burning shell, which dumps more helium into the core as it burns outward through the star.

THE HORIZONTAL BRANCH

On the horizontal branch, the star burns helium in the core, and hydrogen in a shell surrounding the core. The stars become hotter and smaller during this phase, and move across the H-R diagram. These stars stay at about the same luminosity because, even though they are hotter, they are smaller. This is called the horizontal branch (Fig. 8-2).

PULSATING STARS

As stars move across the horizontal branch, some of them begin to pulsate. These pulsating or variable stars change in size, growing bigger and smaller repeatedly, and their luminosity changes in a periodic way. Pulsating stars are not in equilibrium. Their outer shells trap energy, heat up, and expand. As they expand they cool off, and are pulled inward by gravity, which increases their pressure. The compressed layer is denser and absorbs energy, heats up, and so on. These stars are found on the **instability strip**, which crosses the horizontal branch (Fig. 8-2).

There are two important types of pulsating stars:

- **Cepheid variables**. These stars make good standard candles, because there is a period–luminosity relationship. Cepheids vary in a unique way (Fig. 8-3), and so they are easily identified. Cepheids with the same period have the same luminosity. If a star in a distant galaxy is identified as a Cepheid by looking at the light curve, then the period can be measured and, from that, the luminosity. The apparent magnitude can also be determined, so the distance to the star can be found using the inverse square law. The distance to galaxies can be determined to fairly good accuracy using Cepheid variables.

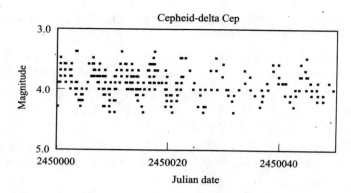

Fig. 8-3. Light curve of a Cepheid variable. (Courtesy of AAVSO.)

- **RR Lyrae stars**. These stars are small and dim, and also have a period–luminosity relationship. Because they are dim, they are not useful for finding distances to other galaxies, but they *are* useful for finding distances to globular clusters (groups of many stars all bound together by gravity) in our own galaxy. These distances have been very cleverly used to find the distance to the center of the Milky Way.

ASYMPTOTIC GIANT BRANCH

Eventually, the star runs out of helium in the core. It moves back across the horizontal branch as the surface temperature falls, and onto the asymptotic giant branch (AGB). While on the asymptotic giant branch, a sequence of changes occur that lead to the eventual ejection of the star's outer layers as a planetary nebula and the formation of a white dwarf star:

1. Helium burning begins in the **shell**, causing the luminosity to increase, while the temperature remains about the same (i.e., the star gets bigger).
2. The star expands, cools, but becomes more luminous.
3. Helium builds up in a shell around the (mostly) carbon core, until there's enough of it at high enough pressure to ignite.
4. The mass of the core increases.
5. A wind develops, and the mass-losing phase begins: the star begins to lose mass at a phenomenal rate, as much as 10^{-5} solar masses (M_{Sun}) per year. At this rate, the entire Sun would disappear in only 10,000 years.
6. This lost mass moves away from the star, and forms a shell around it, which is opaque to visible and ultraviolet light. These shells are so cool that they are most easily observed in the infrared.

CHAPTER 8 Stellar Evolution

WHITE DWARFS

As the mass leaves the star, it exposes hotter and hotter depths. The central star made of carbon (ash of helium fusion) and helium gets bluer and bluer, until it is emitting in mostly the UV. This UV light ionizes the dust and gas that was ejected, which then glows. The dust and gas is a planetary nebula, surrounding a **white dwarf**. The Ring Nebula (Fig. 8-4) is a perfect example of a textbook planetary nebula. A white dwarf is a small, hot star of less than 1.4 solar masses—stellar cores more massive than this will not become white dwarfs, but will instead continue to shrink to become either neutron stars or black holes. White dwarfs, neutron stars, and black holes will be discussed in more detail in Chapter 9.

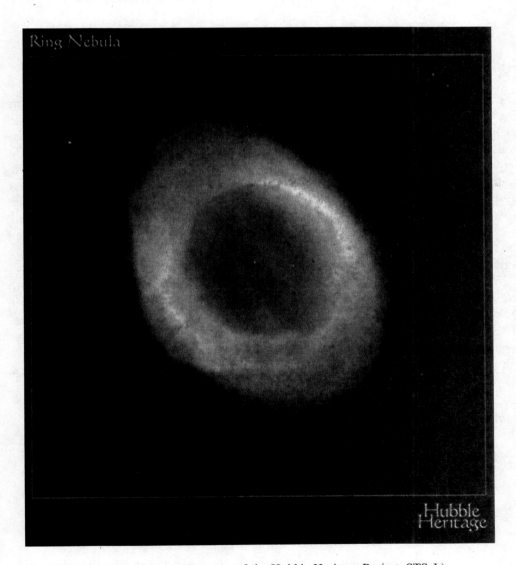

Fig. 8-4. The Ring Nebula. (Courtesy of the Hubble Heritage Project, STScI.)

Stars $> 8M_{Sun}$: Supernovae

If a star is very massive, it will not only evolve faster but will also manufacture heavier elements than lower mass stars do (Fig. 8-5). These stars will go through a mass-losing phase, and then finally explode as supernovae. There are two types of supernovae: Type I and Type II. A Type II supernova is the end of a star's life, and happens to isolated massive stars. A Type I supernova occurs in a binary system, and is not a part of the star's evolution, but is caused by the interaction of the pair of stars. Depending on the spectral details, we have Type Ia, Ib, and Ic supernovae. Observable supernovae should happen about once each century in our galaxy. We have not seen any in our galaxy in modern times. This is because of interstellar extinction due to dust in the Galaxy (see Chapter 10), which prevents us from observing the entire Galaxy.

TYPE II SUPERNOVAE

The internal structure of an evolved massive star looks something like an onion. The outer layer is a non-burning hydrogen envelope. Inside of this are a hydrogen-burning shell, then a helium-burning shell, then carbon, then oxygen, neon, magnesium, and silicon. In the center of it all is a core of iron ash. In this core, the density, pressure, and temperature in the core of the massive evolved star eventually become high enough that the iron begins to fuse. But iron is special. It is the lightest element for which fusion absorbs energy rather than releasing it. In other words, it takes more energy to cause iron to fuse than comes out in the reaction.

Once the iron ignites, the gas will cool, because the fusing iron takes heat out of the system. As the temperature falls, so does the pressure, and the core begins to collapse. The collapsing core drives the temperature still higher, to 10 billion K.

When the temperature reaches 10 billion K, the photons have enough energy to tear the atoms apart in a process called **photo-disintegration**. All of the elements in the core—helium, carbon, oxygen, iron, etc.—are destroyed. Over 10 billion years of fusion activity is reversed in less than 1 second. Like the burning of iron, this process absorbs energy, so that the core can collapse even more. The core is now composed of only protons, neutrons, electrons, and light. As the core collapses, the density rises.

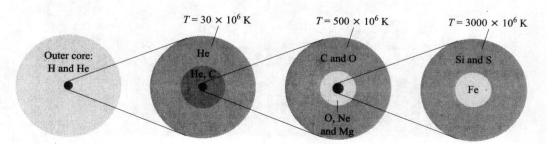

Fig. 8-5. A cutaway of an evolved massive star with heavy elements inside.

CHAPTER 8 Stellar Evolution

When the density reaches $\sim 10^{12}$ kg/m^3, the protons and the electrons become fused into neutrons and neutrinos. These neutrinos have no difficulty in leaving the core of the star, and carrying away energy into the Universe.

Now that there is no charge, the core can collapse even further, until the density surpasses 10^{15} kg/m^3. The neutrons are degenerate at these densities, and resist being packed any more closely together. However, because the infalling material has gathered momentum, it overshoots the degenerate density, and the core density rises to as much as 10^{17}–10^{18} kg/m^3.

Just like a ball hitting a wall, the material in the core rebounds from the dense, degenerate center. The rebound produces shock waves, which travel outwards at high speed, picking up material as they go. These shock waves carry material from the envelope away from the core. In other words, the star explodes.

The entire process, from start of collapse to rebound, takes less than 1 second. For a few days, the supernova may be as bright as a whole galaxy of billions of stars. The equivalent of the entire lifetime energy budget of the Sun is released in this explosion—and that's just in light energy. 100 times as much energy is carried by neutrinos.

TYPE I SUPERNOVAE

An evolved binary star system, in which the white dwarf star is re-accreting mass, ignites all of the carbon at once and blows the white dwarf to bits, producing an explosion of supernova brightness.

SUPERNOVA REMNANTS

The material lost from a supernova moves outward from the central star in much the same way that material moves away from an asymptotic giant branch star. In the case of a supernova, the resultant nebula is called a **supernova remnant**. The Crab Nebula (Fig. 8-6) was created by a supernova observed on Earth in AD 1054. For nearly a month, this supernova was bright enough to see in the daytime. Figure 8-6 shows what is left "now." (Since the nebula is about 1,800 pc away, this image actually shows what the nebula looked like 5,400 years ago.) Doppler shifts of the gas indicate that the Crab Nebula is expanding at a rate of several thousand km/s.

Where Do We Come From?

There are 112 different elements in the current version of the periodic table. Of these elements, 91 occur naturally on the Earth: 81 of these are stable (not radioactive) elements; 10 are radioactive elements, and therefore decay. All the everyday objects with which you interact are composed of a subset of these 112 elements.

Current particle abundances in the Universe are about:

Fig. 8-6. The Crab Nebula. (Courtesy of European Southern Observatory.)

1. Hydrogen: 90%
2. Helium: 9%
3. Li, Be, B: 0.000001%
4. C, N, O, F, Ne: 0.2%
5. Si–Mn: 0.01%
6. Fe–Ge: 0.01%
7. Midweights: 0.00000001% = 1×10^{-7}%
8. Heavyweights: 0.000000001% = 1×10^{-8}%

CHAPTER 8 Stellar Evolution

There are four different processes involved in the production of elements, and each element is produced via some combination of these processes.

1. **Primordial elements**. In the beginning, just after the Big Bang (see Chapter 11), there were only hydrogen and helium atoms in the Universe. By the time the Universe had cooled off enough for the helium atoms to survive the intense heat, the Universe was too cool to manufacture anything heavier.
2. **Stellar nucleosynthesis** (p–p chain, helium fusion, carbon fusion, etc.) produced all the elements up to iron.
3. **Neutron capture (s-process)**. Elements heavier than iron can be formed from iron, by a process called neutron capture. Neutrons get absorbed by the nucleus of iron. This makes a heavy isotope (an **isotope** is an atom with more or less neutrons than the most abundant form of that atom) of iron. These isotopes are unstable, and decay, with the neutrons turning into protons by emission of an electron and a neutrino. The new element can, in turn, capture neutrons, grow heavier, and then decay. This is called the "s-process" (slow process). It happens in a regular star, and can produce atoms as heavy as bismuth.
4. **Neutron capture (r-process)**. In a supernova explosion, there are many many neutrons all moving with high energy. Any element present is likely to be completely bombarded with neutrons, and to capture many of them. This is how we get elements that are heavier than bismuth. It is called the "r-process" (rapid process).

How do we know that this is how the elements formed?

1. Neutron capture and nuclear decay are well-studied phenomena in laboratory experiments. When the results of these experiments are combined with models of the Universe since the Big Bang, they agree quite well with observation.
2. Technetium-99 is direct evidence that heavy-element formation happens in stars. This element has a half-life of 200,000 years, an astronomically short period of time. But, we see this element in stars today, which means that it must be being created all the time, else it would have all decayed already.
3. Light curves of Type I supernovae look just like the sum of the decay curves of nickel-56 (half-life = 55 days) and carbon-56 (half-life = 78 days), implying that many of these isotopes are produced in the explosion, and subsequently decay.

CHAPTER 8 Stellar Evolution

Solved Problems

8.1. What event signals the end of the main-sequence life of a star?

Stars evolve off the main sequence when they run out of hydrogen in the core. At this point, they must find another way of producing energy. Usually, they begin fusing helium in the core, and hydrogen in a shell around the core.

8.2. Why does the process of creating neutrons from protons and electrons reduce the ability of a giant star to support its own weight?

Protons all have the same charge, and so resist being close together. Electrons also all have the same charge, and resist being packed tightly. Neutrons, on the other hand, have no charge, and so are able to be more densely packed.

8.3. Main-sequence stars of 5 solar masses are thought to evolve to white dwarfs of 1 solar mass. What happened to the other 4 solar masses?

Five solar mass stars become planetary nebulae surrounding white dwarfs. The other 4 solar masses, which are not part of the white dwarf, become the planetary nebula.

8.4. What physical property determines the ultimate fate of a star?

The entire fate of a star, from main-sequence lifetime, to eventual demise, is governed by the mass of the star. The initial composition plays a small part in the length of the lifetime as well, but whether the star explodes as a supernova or as a planetary nebula is governed entirely by its mass.

8.5. Why is an iron core unable to support a star?

When iron fuses, the process absorbs energy. With no energy source in the core of a star, it is unable to support itself against gravity.

8.6. Suppose that asymptotic giant branch stars did not lose mass. What effect would this have on the number of white dwarfs in the Galaxy?

White dwarfs are formed when an asymptotic giant branch star loses mass, ejecting a planetary nebula, and exposing a hot central core. If these stars did not shed mass, there would be far fewer white dwarfs in the Galaxy, but instead, there would be more supernovae, neutron stars, and black holes, depending on the initial mass of the asymptotic giant branch star.

8.7. Only a small percentage of the energy of a Type II supernova is carried away by light and the expansion of the supernova remnant. What happens to the rest of the energy? What fraction of the total energy is disposed of in this manner?

The vast majority of energy in a supernova is carried away by neutrinos. As stated in the text, 100 times as much energy is carried away by neutrinos as by light and expansion of the remnant. There are a total of 101 parts of energy (the neutrino part plus the light/expansion

CHAPTER 8 Stellar Evolution

part), of which 100 are carried away by neutrinos. So the fraction of energy carried away by neutrinos is

$$\text{fraction} = \frac{100}{101}$$
$$\text{fraction} = 0.99$$

or 99%.

8.8. Make a flow chart (Fig. 8-7) that shows how a 4 solar mass star evolves after leaving the main sequence.

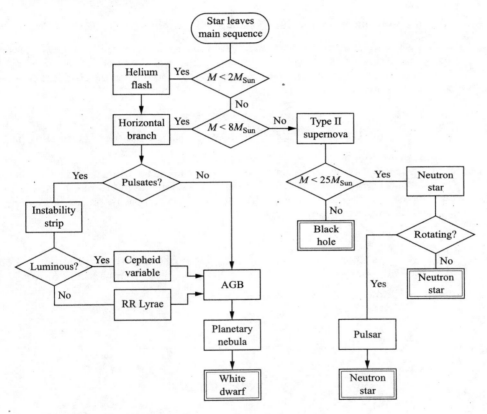

Fig. 8-7. Flowchart of post-main-sequence evolution.

8.9. Why are most of the stars in the H-R diagram on the main sequence?

Stars spend the majority of their lives on the main sequence, so the majority of stars can be found there.

8.10. How does the luminosity of a giant star compare with the luminosity of a main-sequence star of the same spectral class?

Since the same spectral class means that the stars have the same temperature, each square meter of surface emits the same amount of energy (and so is the same brightness). However, the giant star is much larger than the main-sequence star: the radius is about 100 times as large. Since the luminosity of the star goes like the surface area, which in turn scales as the radius squared, the giant star is $100^2 = 10,000$ times as bright as the main-sequence star.

8.11. Why do nuclei of elements other than hydrogen require higher temperatures to fuse?

Hydrogen has the fewest number of protons of all elements. The repulsion between protons is the main impediment to fusion. Therefore elements with more protons will have to be traveling faster to overcome this repulsion. In a gas, faster atoms mean higher temperatures.

8.12. Why does chemical composition change most rapidly in the center of a star?

In the center of a star, the temperatures, densities, and pressures are the highest. This means that atoms interact more often, which increases the probability of fusion occurring. More atoms interact each second, and more atoms fuse each second, so the chemical composition changes most rapidly in the core.

8.13. What effect do supernova explosions have on the makeup of interstellar gas?

Supernova explosions eject lots of processed material in the remnant. As this remnant expands, the density decreases until the remnant cannot be distinguished from the rest of the interstellar medium. All of the processed material remains in the interstellar medium. So the supernova explosions "enrich" the interstellar medium with processed material, i.e., heavier elements than hydrogen and helium.

8.14. Calculate the escape velocity for a 1 M_{Sun} giant of radius 1 AU. Planetary nebulae are observed to expand at 10–50 km/s. How do these numbers compare?

Recall the escape velocity equation from Chapter 1:

$$v_e = \sqrt{\frac{2GM}{d}}$$

$$v_e = \sqrt{\frac{2 \cdot 6.67 \times 10^{-11} \frac{m^3}{kg \cdot s^2} \cdot 2 \times 10^{30} \, kg}{1.5 \times 10^{11} \, m}}$$

$$v_e = 42{,}000 \, m/s$$

$$v_e = 42 \, km/s$$

This number is at the high end of the planetary nebula expansion range, but still within the observed range of velocities. Larger stars will have lower escape velocities, while more massive ones will have higher escape velocities.

8.15. The Crab Nebula is about 230 arcseconds across. What is its linear diameter?

From Chapter 1, the relationship between angular size, distance, and linear size is

$$\theta('') = 206{,}265 \cdot \frac{d}{D}$$

CHAPTER 8 Stellar Evolution

From page 163, the distance to the Crab Nebula is 1,800 pc. We can use this fact, and rearrange the equation to find the linear size:

$$d = \frac{\theta(") \cdot D}{206{,}265}$$
$$d = \frac{230 \cdot 1{,}800}{206{,}265}$$
$$d = 2 \, \text{pc}$$

8.16. The Crab Nebula is expanding at about 1,400 km/s. Calculate its age. How does this compare to the supernova date given in the text?

The radius of the Crab Nebula is about 1 pc, or 3×10^{13} km. Use the relationship between travel time, distance, and velocity to find out how long ago the material left the central star:

$$t = \frac{d}{v}$$
$$t = \frac{3 \times 10^{13} \, \text{km}}{1{,}400 \, \text{km/s}}$$
$$t = 2 \times 10^{10} \, \text{s}$$

This is not a very useful number. Convert it to years by dividing by 3.16×10^7 s/yr. The age of the nebula is about 700 yr. This would place the supernova at about AD 1300, nearly 300 years from the "proper" date.

8.17. Find the density of a red giant star of 1 M_{Sun}, radius 1 AU. How does this compare with the density of air (1 kg/m³)?

Assume the star is perfectly spherical. The density is given by the mass divided by the volume:

$$\rho = \frac{M}{V}$$
$$\rho = \frac{M}{\frac{4}{3}\pi \cdot R^3}$$
$$\rho = \frac{2 \times 10^{30} \, \text{kg}}{\frac{4}{3}\pi \cdot (1.5 \times 10^{11} \, \text{m})^3}$$
$$\rho = 0.00014 \, \text{kg/m}^3$$

This is far less than the density of air.

8.18. Globular clusters are groups of stars all born at the same time out of the same cloud. A representative sample of the stars in an imaginary globular cluster are plotted on the H-R diagram shown in Fig. 8-8, along with a bar indicating the main-sequence positions of stars of certain masses. How old is this globular cluster?

Since the highest mass star in the globular cluster still on the main sequence is about 2 M_{Sun}, the age of the globular cluster must be the main-sequence lifetime of a 2 M_{Sun} star:

$$\frac{t}{t_{Sun}} = \left(\frac{M}{M_{Sun}}\right)^{-2.5}$$

$$t = \left(\frac{M}{M_{Sun}}\right)^{-2.5} \cdot t_{Sun}$$

$$t = (2)^{-2.5} \cdot 10 \text{ billion years}$$

$$t = 1.8 \text{ billion years}$$

This imaginary globular cluster is about 2 billion years old.

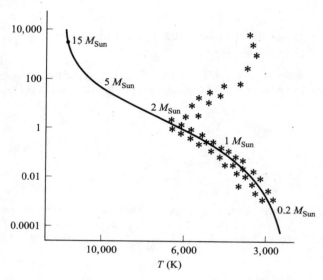

Fig. 8-8. H-R diagram of a globular cluster with the main-sequence locations of stars of various masses indicated.

Supplementary Problems

8.19. How much gravitational energy does the Sun have now?

Ans. 3.83×10^{31} J

8.20. How much gravitational energy does the Earth have now?

Ans. 3.73×10^{32} J

8.21. How much mass is converted into energy each time four hydrogen nuclei become a helium nucleus?

Ans. 4.75×10^{-29} kg

CHAPTER 8 Stellar Evolution

8.22. Suppose that a pulsating star has a change in magnitude of 1 from maximum brightness to minimum brightness. What is the ratio of the radii at maximum and minimum?

Ans. 2.5

8.23. What is the escape velocity from the surface of a white dwarf, $M = 1\,M_{Sun}$, $R = 0.9\,R_{Earth}$?

Ans. 6.8×10^6 m/s

8.24. Suppose a particular supernova increases by 6 magnitudes when it explodes. What is the corresponding increase in brightness?

Ans. Brightness increases by a factor of 244

8.25. A nebula is observed in to have a line of rest wavelength 989.06 nm blue-shifted to 988.96 nm. Is this a supernova or a planetary nebula?

Ans. Planetary nebula

8.26. In the center of a red giant undergoing the helium flash, what is the peak wavelength of the blackbody radiation from the core?

Ans. 9.7×10^{-12}

8.27. What is the total luminosity (in watts) of a supernova that is as bright as a billion Suns?

Ans. 3.8×10^{35} W

8.28. Suppose the inner 1 M_{Sun} of a Type II supernova collapses to a density of 10^{17} kg/m^3. What is the radius of this core?

Ans. 17,000 m

8.29. Suppose the entire mass of the Sun were converted into energy. How much energy would be produced?

Ans. 1.8×10^{47} J

8.30. A star is plotted in the upper right corner of the H-R diagram. What kind of star is this?

Ans. Red giant

8.31. A star is plotted exactly in the middle of an H-R diagram. What is the color? What is the luminosity class?

Ans. Yellow, dwarf (V)

CHAPTER 9

Stellar Remnants (White Dwarfs, Neutron Stars, and Black Holes)

Degenerate Gas Pressure

In an ordinary gas, the pressure depends on the density and on the temperature. At very high densities, a mutual repulsion develops between electrons. This repulsion is not due to the classical behavior of their electrical charge: rather, it is due to their quantum mechanical properties. This repulsion produces an additional pressure, the so-called degenerate pressure, which depends on the density alone, not the temperature. Thus, the material can be heated without expanding, and can be cooled without shrinking. The degenerate pressure halts the gravitational collapse, like the ideal gas pressure, with one significant difference: when material is added, the gravity of the star increases, but the increase in the degenerate pressure is not as high as in ordinary matter. Therefore, the star shrinks. The higher the mass of a degenerate star, the smaller its volume.

Electrons are degenerate when $\rho = 10^9 \, \text{kg/m}^3$. This is about the same as squeezing a house into the cap of a soda bottle. Neutrons are degenerate when $\rho = 10^{18} \, \text{kg/m}^3$. This is about the same as squeezing the entire Earth into a box that is only 200 meters on each side! Most stars end their lives as small dense spheres of degenerate matter that slowly cool and radiate heat. Because the surface is small, the cooling is slow and could take longer than the current age of the Universe.

White Dwarfs

White dwarfs are hot (~10,000 K), low luminosity stars composed mainly of carbon and helium. Their luminosity is low because their surface is small. The electrons in white dwarfs are degenerate. As mentioned above the more mass in the white dwarf, the smaller the radius. This peculiar mass–radius relation is shown in Table 9-1.

Table 9-1. Mass–radius relationship for white dwarfs

Mass	Radius
0.5 M_{Sun}	1.5 R_{Earth}
1.0 M_{Sun}	0.9 R_{Earth}
1.3 M_{Sun}	0.4 R_{Earth}

The largest possible white dwarf mass is 1.4 M_{Sun}. This is called the Chandrasekhar limit, and is the most mass that the electron degenerate core can support without collapsing under its own gravity.

Once a white dwarf has collapsed, it cools off, but the degenerate pressure inside does not drop and the star does not shrink on cooling. As the star cools, its luminosity decreases, just as you would expect. This process slows over time, so that the dimmer the star gets, the slower it gets dimmer. Table 9-2 shows the age–luminosity relationship for a 0.6 M_{Sun} white dwarf.

Table 9-2. Age–luminosity relationship for white dwarfs

Age	Luminosity
20 million yr	0.1 L_{Sun}
300 million yr	0.01 L_{Sun}
1 billion yr	0.001 L_{Sun}
6 billion yr	0.0001 L_{Sun}

At 6 billion years, the temperature (and therefore the color) of the white dwarf is about the same as the Sun's surface, but the luminosity is much less, because the star is so small.

White dwarfs are the central stars of planetary nebulae, and their final mass depends on their initial main-sequence mass, as shown in Table 9-3.

CHAPTER 9 Stellar Remnants

Table 9-3. The white dwarf mass depends on the main-sequence mass

Main-sequence mass	White dwarf mass
2–8 M_{Sun}	0.7–1.4 M_{Sun}
< 2 M_{Sun}	0.6–0.7 M_{Sun}
< 1 M_{Sun}	< 0.6 M_{Sun}

No white dwarfs with masses less than 0.6 M_{Sun} have been observed. The Universe may not be old enough for stars of less than 1 M_{Sun} to have evolved to the white dwarf stage.

Solved Problems

9.1. Suppose that two white dwarfs have the same surface temperature. Why is the more massive of the two white dwarfs less luminous than the less massive one?

White dwarfs have the peculiar property that more massive ones are smaller. Therefore, the more massive white dwarf has a lower luminosity because there is less surface area to emit light.

9.2. Where are white dwarfs located on an H-R diagram?

Because white dwarfs are small, they would be located near the bottom. Because they are white (i.e., hot), they would be located to the left. White dwarfs should be found below and to the left of main-sequence stars on the H-R diagram.

9.3. What is the density of a 1 solar mass (1 M_{Sun}) white dwarf?

From Table 9-1, the radius of a 1 M_{Sun} white dwarf is $0.9 R_{Earth}$ or $0.96 \times 6,378,000\,m = 5,740,000\,m$. The density is given by

$$\rho = \frac{M}{V}$$
$$\rho = \frac{M}{\frac{4}{3}\pi \cdot R^3}$$
$$\rho = \frac{2 \times 10^{30}\,kg}{\frac{4}{3}\pi \cdot (5.74 \times 10^6\,m)^3}$$
$$\rho = 2.5 \times 10^9\,kg/m^3.$$

The density of a 1 M_{Sun} white dwarf is 2.5×10^9 kg/m³. This is consistent with it being composed of electron degenerate matter. Electrons are degenerate when densities are greater than 10^9 kg/m³.

Neutron Stars

Main-sequence stars with mass in the range of 8–25 M_{Sun} become supernovae, lose a lot of their envelope, and leave a degenerate neutron core behind. Like white dwarfs, neutron stars also get smaller when they are more massive. For a sense of scale, a 0.7 M_{Sun} neutron star has a radius of ~10 km (about 6.25 miles). The maximum mass of a neutron star is probably somewhere between 1.5 and 2.7 M_{Sun}.

Neutron stars collapse from larger stars, and during that collapse, angular momentum must be conserved. Therefore, the stars are rotating very rapidly, about 1,000 times per second! They also have large magnetic fields, due to the collapse and compression of the original magnetic field of the star.

One special type of neutron star is the **pulsar**. Pulsars are spinning neutron stars, with the magnetic field axis not aligned with the rotation axis. Along the magnetic field lines that come out of the pole are many spiraling charged particles, which emit radiation. When one of these poles is pointed towards us, we detect a flash of radio waves. When neither is pointed towards us, we don't detect any radio waves. The whole system functions very much like a lighthouse. Figure 9-1 shows a diagram of this model.

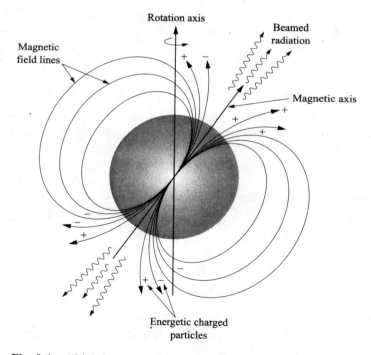

Fig. 9-1. "Lighthouse model" of a pulsar.

CHAPTER 9 Stellar Remnants

The pulses from a pulsar sometimes vary: sometimes individual pulses are shaped differently, and sometimes there are quiet periods with no pulses at all.

The spinning of the pulsar slows over time. It takes about 30 million years for the rotation to slow from 0.001 seconds to 2 seconds. This is short compared with the age of the Universe, so you might expect to see lots of these with periods longer than 2 seconds. However, no pulsar has ever been observed with a period longer than 5 seconds. This implies that there must be some sort of shut-off mechanism for the charged particles as the star slows its spinning.

Solved Problems

9.4. Why do neutron stars rotate more rapidly than main-sequence stars?

Neutron stars were once main-sequence stars. When they contracted, the angular momentum was conserved. Since the radius decreased, the velocity must increase in order for the total angular momentum to remain constant ($L = mvr$).

9.5. Why would no pulses be observed from a rotating neutron star if the magnetic field axis and rotation axis were aligned?

If these two axes were aligned, the magnetic field axis would always point in the same direction. We would either observe radio waves all the time, if the magnetic axis were pointed in our direction, or no radio waves, if it were pointed away from us.

9.6. The angular momentum of an object is conserved. Assuming all the mass of a stellar core is retained during collapse, how fast was the progenitor of a 10 km pulsar spinning when it was one solar radius ($\times 10^5$ km)? (Assume a rotation period of 0.001 seconds.)

$$L = m_1 v_1 r_1 = m_2 v_2 r_2$$

But $v = \omega r$

$$m_1 \omega_1 r_1^2 = m_2 \omega_2 r_2^2$$

Since the mass does not change,

$$\omega_1 r_1^2 = \omega_2 r_2^2$$
$$\omega_1 = \frac{\omega_2 r_2^2}{r_1^2}$$

The period, P, is related to the angular velocity ω by $\omega = 2\pi/P$, so

$$\omega_1 = \frac{2\pi \cdot r_2^2}{r_1^2 \cdot P_2}$$

$$\omega_1 = \frac{2\pi \cdot 10^2}{(7 \times 10^5)^2 \cdot 0.001}$$

$$\omega_1 = 12.28 \times 10^{-7}/\text{s}$$

Find the period, P_1:

$$P_1 = 4.9 \times 10^6 \, \text{s}$$

$$P_1 = 57 \, \text{days}$$

The progenitor was spinning once every 57 days. This is half as fast as the rotation of the Sun, which has a rotation period of 25 days.

9.7. If you were compressed to neutron star densities, how large would you be?

Assume a mass of 65 kg. The density of a neutron star is the density of neutron-degenerate matter, about 10^{18} kg/m^3. The relationship between density, volume, and mass is

$$\rho = \frac{M}{V}$$

Rearranging this equation to solve for volume, and plugging in the above values gives

$$V = \frac{M}{\rho}$$

$$V = \frac{65}{10^{18}} \, \text{m}^3$$

$$V = 6.5 \times 10^{-17} \, \text{m}^3$$

If you were compressed to neutron star density, you would occupy a cube measuring 0.000004 m on each side, since $0.000004 \times 0.000004 \times 0.000004 = 6.5 \times 10^{-17}$.

9.8. What is the escape speed of a 0.5 solar mass white dwarf?

Table 9-1 shows that the radius of a $0.5 \, M_{\text{Sun}}$ white dwarf is $1.5 \, R_{\text{Earth}}$, or $1.5 \cdot 6.378 \times 10^6 \, \text{m} = 9.567 \times 10^6 \, \text{m}$. Recall the escape velocity equation from Chapter 1,

$$v_e = \sqrt{\frac{2GM}{d}}$$

$$v_e = \sqrt{\frac{2 \cdot 6.67 \times 10^{-11} \, \text{m}^3/\text{kg}/\text{s}^2 \cdot 1 \times 10^{30} \, \text{kg}}{9.567 \times 10^6 \, \text{m}}}$$

$$v_e = 3.7 \times 10^6 \, \text{m/s}.$$

This is only about 80 times less than the speed of light. Most ordinary things, such as rocket ships or space shuttles, could not even come close to escaping the surface of a white dwarf. Light, however, easily escapes.

9.9. Suppose you were standing on the equator of a neutron star (radius 10 km, rotation rate 0.001 seconds). How fast would you be traveling (in km/s)? How does this compare with the speed of light?

CHAPTER 9 Stellar Remnants

A rotation rate of 0.001 seconds means that you are traveling around an entire circle of radius 10 km in only 0.001 seconds. The circumference of a circle is

$$C = 2\pi \cdot r$$
$$C = 2\pi \cdot 10\,\text{km}$$
$$C = 62.8\,\text{km}$$

So you travel 62.8 km in 0.001 seconds, or are traveling 62,800 km/s. The speed of light is 300,000 km/s, so you would be traveling at approximately 20% of the speed of light!

Black Holes

Main-sequence stars with masses larger than 25 M_{Sun} become supernovae, leaving behind a core that is more massive than the neutron star, about 3 M_{Sun} or more. Therefore the gravity is higher, and can overcome the degenerate neutron pressure, and the mass collapses to a black hole. The laws that apply inside a black hole remain unknown, yet it is clear that the escape velocity from the vicinity of these extremely dense objects must be very high, and possibly higher than the speed of light. Recalling the escape velocity equation (Chapter 1), and setting $v = c$, we have

$$c = \sqrt{\frac{2GM}{d}}$$

and rearranging, we have

$$d = \frac{2GM}{c^2}$$

Within this distance (the **Schwarzschild radius**), nothing can escape the gravitational influence of the black hole.

HAVE BLACK HOLES BEEN OBSERVED?

We have seen stars in binary systems, where we can observe only one star, and it behaves as though it has a companion of a few solar masses. These systems are also strong sources of X-rays. There are about half a dozen of these in our galaxy, and these are possibly locations of remnants of massive stars which have become black holes.

We observe the centers of galaxies (including our own) as huge sources of X-rays. Astronomers can observe the speed of rotation near the center (from the Doppler shift), and therefore determine the mass within the orbits (from the circular velocity equation). This density is too high to be anything but a black hole.

CHAPTER 9 Stellar Remnants

Solved Problems

9.10. If the Sun were suddenly replaced by a one solar mass (1 M_{Sun}) black hole, what would happen to the solar system?

The solar system would become quite dark and cold, but otherwise, nothing would happen. The planets would still revolve and rotate around the Sun with the same periods as they currently possess.

9.11. If the Earth were suddenly replaced by a black hole with the mass of the Earth, what would happen to the Moon? What if the black hole had the mass of the Sun?

If the Earth were replaced by a black hole with the mass of the Earth, nothing would happen to the Moon. It would orbit the Earth with the same period it currently has.
   However, if the black hole had the mass of the Sun, the increased gravitational attraction would pull the Moon closer. Would the Moon fall into the black hole? To answer that, we need to compare the Schwarzschild radius of a one solar mass black hole with the radius of the orbit of the Moon (3.84×10^8 m).

$$R_s = \frac{2GM}{c^2}$$
$$R_s = \frac{2 \cdot (6.67 \times 10^{-11} \, m^3/kg/s^2) \cdot (2 \times 10^{30} \, kg)}{(3 \times 10^8 \, m/s)^2}$$
$$R_s = 3000 \, m$$

This is vastly smaller than the distance from the Earth to the Moon, so the Moon would not be "swallowed up" by the black hole, although its orbital radius would decrease somewhat.

9.12. In general relativity, space is bent by the presence of a mass. One experiment that supported this theory showed that stars appear farther apart when viewed towards the Sun than when viewed away from the Sun. How would the results of this experiment change if space were not bent by the presence of mass? How would the results change if the space were negatively curved (so it made a "hill" instead of a "hole")?

If space were not curved by mass, the positions of the stars would be unchanged no matter where they were viewed. A picture of the background stars during an eclipse could be placed on top of a picture of the stars 6 months later, and they would line up exactly.
   If space were negatively curved, so the mass made a "hill" instead of a "hole," the stars would appear to move **closer** to the Sun during the eclipse. (The light would roll "down" the hill.) So the stars would appear closer together instead of farther apart.

9.13. What is the Schwarzschild radius of a black hole of one million (1×10^6) solar masses? one billion (1×10^9) solar masses? How do these radii compare to the size of the solar system?

An object of one million solar masses has a mass of 2×10^{36} kg. An object of one billion solar masses has a mass of 2×10^{39} kg. So the Schwarzschild radius of the one million solar mass black hole is

CHAPTER 9   Stellar Remnants

$$R_s = \frac{2GM}{c^2}$$

$$R_s = \frac{2 \cdot 6.67 \times 10^{-11} \, \text{m}^3/\text{kg/s}^2 \cdot 2 \times 10^{36} \, \text{kg}}{(3 \times 10^8 \, \text{m})^2}$$

$$R_s = 3 \times 10^9 \, \text{m}.$$

This is smaller than the orbit of the Earth (1 AU = 1.5 × 10^{11} m).

Since the 1 billion solar mass black hole is 1,000 times as massive as the 1 million solar mass black hole, its Schwarzschild radius is also 1,000 times as large, or 3×10^{12} m. This is larger than the orbital radius of Uranus (2.87×10^{12} m), but smaller than the orbital radius of Neptune (4.4×10^{12} m).

9.14. Every object has an event horizon. One way to define a black hole is to say that it is an object that is smaller than its own event horizon. How small would you have to be in order to become a black hole? How does this compare with the size of an atomic nucleus (radius = 10^{-15} m)?

Assume a mass of 65 kg. The Schwarzschild radius of a 65 kg black hole is

$$R_s = \frac{2GM}{c^2}$$

$$R_s = \frac{2 \cdot 6.67 \times 10^{-11} \, \text{m}^3/\text{kg/s}^2 \cdot 65 \, \text{kg}}{(3 \times 10^8 \, \text{m/s})^2}$$

$$R_s = 9.6 \times 10^{-26} \, \text{m}$$

This is **much** smaller than the radius of a single atom.

9.15. A black hole has a Schwarzschild radius of 30 km. How much mass is contained in the black hole?

The Schwarzschild radius equation can be rearranged to solve for the mass,

$$M = \frac{R_s \cdot c^2}{2G}$$

$$M = \frac{30{,}000 \, \text{m} \cdot (3 \times 10^8 \, \text{m/s})^2}{2 \cdot 6.67 \times 10^{-11} \, \text{m}^3/\text{kg/s}^2}$$

$$M = 2 \times 10^{31} \, \text{kg}$$

The mass of this black hole is 2×10^{31} kg, or 10 times the mass of the Sun.

Supplementary Problems

9.16. What is the ratio of the volume of a 0.7 M_{Sun} neutron star to the volume of a 0.7 M_{Sun} white dwarf?

Ans. 1×10^{-9} (1 billionth)

CHAPTER 9 Stellar Remnants

9.17. What is the radius of the event horizon of a black hole of mass 100 M_{Sun}?

Ans. 300 km

9.18. What is the escape speed of a 0.7 M_{Sun} neutron star?

Ans. 1.4×10^8 m/s

9.19. You observe a pulsar with a period of 2 seconds. How long has this star been a neutron star?

Ans. 30 million years

9.20. What is the mass of a black hole with Schwarzschild radius 1.0×10^2 km?

Ans. 340 M_{Sun}

9.21. What is the peak wavelength of emission from a white dwarf?

Ans. 2.9×10^{-7} m

9.22. What is the escape speed from the Schwarzschild radius of a 100,000 M_{Sun} black hole?

Ans. c

9.23. Compare the answer to Problem 9.22 to the speed of light.

Ans. It is the same

CHAPTER 10

Galaxies and Clusters

The Milky Way

The Milky Way is our own galaxy. Since we are a part of the Milky Way, it is difficult to construct an external view. Many parts of the Galaxy are not observable. However, through various methods, we have determined the size and shape of our Galaxy, and our place in it.

THE CENTER

In 1915, Shapley attempted to figure out the location of the center of the galaxy by mapping out the positions of globular clusters. A globular cluster is a group of hundreds of thousands of stars, all tightly bound together in a ball 20–100 pc across. They have a very distinctive appearance, and so are easy to find.

He discovered that 20% of the clusters lay in the direction of the constellation Sagittarius, and occupied only about 2% of the sky, and inferred that the bulk of the Galaxy must lie in that direction. He used RR Lyrae stars (see Chapter 8) to estimate the distances to the globular clusters. He found that these globular clusters are centered on a point about 15 kpc from the Sun. Shapley reasoned that if the clusters were distributed evenly about the Galaxy (and there's no reason to think they shouldn't be), then the center of the Galaxy lies in the direction of Sagittarius, about 15 kpc away. Shapley didn't know about dust, so he overestimated the distance to the center of the Galaxy by a factor of two. The center is actually about 8.5 kpc away.

THE SHAPE

We think that the Milky Way Galaxy is a spiral galaxy, composed of three parts:

1. *The disk*. Most of the stars in the Milky Way are located in a disk about 2 kpc thick, and about 40 kpc across. The disk is composed primarily of hot, young stars, and contains lots of dust and gas.

2. *The bulge*. The bulge lies in the center of the Galaxy. It is about 6 kpc across, and extends above and below the disk. There is very little gas or dust (the material required to form new stars). Therefore the bulge consists mainly of old stars.

3. *The halo*. The halo is primarily made up of globular clusters, and is roughly spherical in shape. It is centered on the center of the galaxy, and is about 30–40 kpc in radius. Again, there is very little dust and gas—the halo is primarily composed of old stars. Interestingly, most of these stars are low-metallicity stars, which means that they have even fewer metals than the Sun.

Figure 10-1 is a schematic diagram of the Milky Way as viewed from the "top," and from the "side." The edge of the Galaxy is ill-defined. It is hard to say where the influence of our Galaxy ends. There are many distant globular clusters, and even a few nearby dwarf galaxies, that seem to orbit the Milky Way. Whether or not these are considered part of the Milky Way is largely subjective.

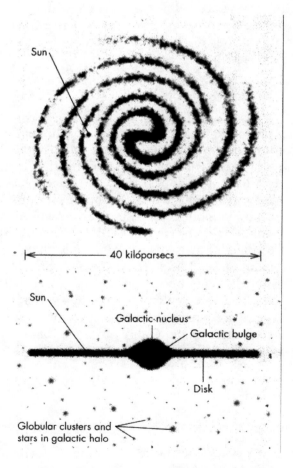

Fig. 10-1. The Milky Way from the top and from the side.

CHAPTER 10 Galaxies and Clusters

ROTATION

The Sun (and everything else in the Galaxy), is in orbit around the center. The Sun's velocity is 220 km/s. The Sun's galactic radius is 8.5 kpc. If we assume that the Sun goes around the center in a circle, we can calculate the amount of time that takes:

$$P = \frac{2\pi \cdot R}{v}$$

where P is the period, R is the radius, and v is the velocity. The orbital period of the Sun around the center of the galaxy is about 240 million years. This is about 5% the age of the Sun. This means that the Sun has been around the center of the Galaxy 20 times already.

We can use Kepler's Third Law to estimate the mass of the Galaxy, M:

$$P^2 = \frac{4\pi^2 a^3}{G(M_{Sun} + M)}$$

$$P^2 = \frac{4\pi^2 a^3}{GM_{Sun}} \cdot \frac{1}{\left(1 + \frac{M}{M_{Sun}}\right)}$$

$$P^2 = \frac{4\pi^2}{GM_{Sun}} \frac{a^3 M_{Sun}}{M}$$

In the last step, we used the fact that $M \gg M_{Sun}$. When measuring P in years and a in AU, the factor

$$\frac{4\pi^2}{GM_{Sun}} = 1$$

and so

$$P^2 = \frac{M_{Sun}}{M} a^3$$

or,

$$M = \frac{a^3}{P^2} M_{Sun}$$

$$M = \frac{(1.8 \times 10^9 \, \text{AU})^3}{(240 \times 10^6 \, \text{yr})^2} M_{Sun}$$

$$M \approx 10^{11} M_{Sun}$$

The galactic rotation velocities and galactic radii of stars can be found from Doppler shifts and their distances from the center. Plotting the rotation velocity vs. galactic radius gives the **rotation curve**. Compare this with a solid object, or with a planetary system, as shown in Fig. 10-2. Near the center, the Galaxy rotates like a solid object. As the distance increases, the velocity does not decrease, as expected for a planetary system. Instead, the rotation curve remains flat. The deviation from planetary motion is in part due to the fact that the mass is not

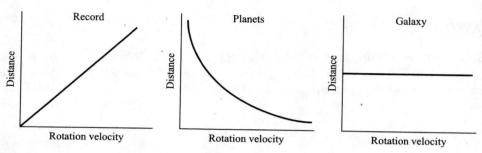

Fig. 10-2. Rotation curves for various kinds of systems.

concentrated at one point, so that Kepler's law does not strictly apply. More importantly, the high velocity of distant objects indicates that the mass of the Milky Way is higher than our estimate of 10^{11} M_{Sun} derived from Kepler's law.

The mass of the entire Milky Way is estimated from these measurements to be about 10^{12} M_{Sun}. However, astronomers cannot see this much matter. All of the stars and planets and dust and so on that we can either observe directly, or infer from its effects on starlight, adds up to only about 10% of the measured mass of the Milky Way: 90% of the Milky Way is made of some kind of mass that we just can't see. We call this "missing mass" **dark matter**.

There are two main ideas about what dark matter might be:

1. *Weakly Interacting Massive Particles* (**WIMPs**). These are theoretical particles much like neutrinos. WIMPs have some mass, but barely interact with the rest of the Universe.

2. *MAssive Compact Halo Objects* (**MACHOs**). These could be objects much like Uranus or Neptune—ordinary matter, but so cold that they don't emit light of their own—or they could be small black holes.

Astronomers are convinced that the majority of the Universe (not just the Galaxy) exists in some as-yet-undetected form.

SPIRAL STRUCTURE

Astronomers believe that the Milky Way is a spiral galaxy, with arms like those in the spiral galaxy NGC4414 shown in Fig. 10-3. Optical observations show hot young stars and HII regions (see Chapter 6) in three big clumps, which offer evidence for three spiral arms in our local vicinity. Out past the nearby spiral arms, optical light is extinguished by dust. Radio telescope observations of the velocities of clouds of neutral hydrogen seem consistent with a spiral arm architecture.

Spiral arms are filled with gas and dust, and therefore are regions of intense star formation. There are lots of young, hot stars (O and B type) in the arms of spiral galaxies, making them brighter and bluer than the surrounding regions.

Spiral arms are actually density waves. These are analogous to bunching up of cars in a highway. If you were to watch this from above, you'd see that the clump

CHAPTER 10 Galaxies and Clusters

Fig. 10-3. Spiral Galaxy NGC4414. (Courtesy of the Hubble Heritage Project, STScI.)

is not composed of the same cars all the time, nor does it move. Instead, cars come up to the jam, slow down, weave their way through, emerge out the other side, and then speed up again. This seems to be the way that spiral arms work. Stars and dust clouds approach the clump, slow down, weave their way through, then speed up again on the other side. This explanation solves one of the big mysteries of spiral arms (Why don't they "wind up"?), by allowing the arms to consist of different stars at different times.

In another model, the spiral arms are the result of supernovae explosions which trigger massive star formation in a chain-reaction pattern. This is the self-propagating star formation model. There are two possible explanations for the formation of the arms.

1. *Tidal influence of companions.* Suppose two elliptical galaxies undergo a near-collision. The stars from each galaxy will be pulled toward the other unevenly across the galaxy, with the near ones being pulled more strongly than the far ones. As the galaxies pass each other, they stretch and twist in response to the changing gravitational forces. This might cause spiral arms to form.

2. *Small irregularities in the disk grow.* Suppose the arms form in response to small over-densities in the disk (places where there are slightly more stars than usual). Since the disk is rotating, these over-dense pockets will orbit the center, and might cause spiral arms to form.

CHAPTER 10 Galaxies and Clusters

Solved Problems

10.1. What is the evidence that most of the mass in the Milky Way lies in some as-yet-undetected form?

If the mass setting the outer parts in orbit was constant, then as the distance increased, the gravitational force and hence the speed, should decrease. The fact that the rotation curve of the Milky Way is flat out to great distances from the center means that the outer parts see a higher mass, which causes them to orbit at a speed higher than expected.

10.2. Using the velocity of the Sun (220 km/s), and its distance from the center of the Milky Way (8,500 pc), calculate how long it takes the Sun to orbit the center.

The total distance the Sun must travel in one orbit is the circumference of a circle of radius 8,500 pc:

$$C = 2\pi \cdot R$$
$$C = 2\pi \cdot 8{,}500 \text{ pc}$$
$$C = 53{,}400 \text{ pc}$$

Multiplying by 3×10^{13} km/pc gives $C = 1.6 \times 10^{18}$ km. At a speed of 220 km/s, the time it will take the Sun to travel this far is

$$t = d/v$$
$$t = \frac{1.60 \times 10^{18} \text{ km}}{220 \text{ km/s}}$$
$$t = 7.28 \times 10^{15} \text{ s}$$
$$t = 7.28 \times 10^{15} \text{ s} \cdot \frac{1 \text{ yr}}{3.16 \times 10^7 \text{ s}}$$
$$t = 2.3 \times 10^8 \text{ yr}$$

The Sun travels once around the center of the galaxy in about 230 million years. This is slightly less than the amount of time quoted in the text, due mainly to round-off error.

10.3. Assume the orbit is circular. How much mass is located within the radius of the Sun's orbit?

Recall the circular velocity equation

$$v_c = \sqrt{\frac{GM}{d}}$$

In this case, v_c and d are known, but M is the unknown variable. Rearrange and substitute.

CHAPTER 10   Galaxies and Clusters

$$M = \frac{d \cdot v_c^2}{G}$$

$$M = \frac{2.6 \times 10^7 \text{ km} \cdot (220 \text{ km/s})^2}{6.67 \times 10^{-11} \text{ m}^3/\text{kg/s}^2}$$

$$M = 2 \times 10^{32} \text{ kg} \cdot \text{km}^3/\text{m}^3$$

$$M = 2 \times 10^{32} \text{ kg} \cdot 10^9$$

$$M = 2 \times 10^{41} \text{ kg}$$

The mass of the galaxy interior to the Sun is about 1.2×10^{42} kg. Compared with the mass of the Sun (2×10^{30} kg), there are roughly 10^{11} (100 billion) solar masses in the part of the Milky Way interior to the Sun.

10.4. Suppose that you observe a star moving at high velocity relative to the Sun in the solar neighborhood. What is the orbit of this star? Where does it spend most of its time? Which direction(s) is it most likely traveling?

The orbit of this star is highly elliptical, and it spends most of its time in the halo. The star is probably traveling into or out of the disk.

10.5. At a distance of 16,000 pc from the center, roughly at the outer edge of the disk, the rotation speed of the Milky Way is 230 km/s. What is the mass of the Milky Way?

16,000 pc is 4.8×10^{20} m. Use the rearranged circular velocity equation:

$$M = \frac{d \cdot v_c^2}{G}$$

$$M = \frac{4.8 \times 10^{20} \text{ m} \cdot (230 \text{ km/s})^2}{6.67 \times 10^{-11} \text{ m}^3/\text{kg/s}^2}$$

$$M = 3.8 \times 10^{35} \text{ kg} \cdot \text{km}^2/\text{m}^2$$

$$M = 3.8 \times 10^{41} \text{ kg}$$

The mass of the entire Milky Way is about 4×10^{41} kg, or about 2×10^{11} solar masses.

Normal Galaxies

There are three types of galaxies that we observe.

1. SPIRALS

Spirals consist of a bulge, a disk (with arms), and a halo. The bulge and the halo are both composed of reddish, older stars, but the disk and the arms are composed of bluish, younger stars, and lots of accompanying dust and gas. The stars in the disk orbit the center of the galaxy in an ordered fashion. All of the stars travel around the bulge in the same "sense" (clockwise or counterclockwise). The stars in the halo and the bulge, however, travel in disordered orbits. These stars can orbit the center in either sense, or with any inclination to the disk.

Spirals are subdivided into classes a, b, and c. These classes roughly indicate the size of the bulge relative to the galaxy. A type Sa galaxy (Spiral a) has a large bulge, and tightly wound arms. A type Sc galaxy (Spiral c) has a small bulge, and very prominent arms. Figure 10-4 shows images of galaxies of each type.

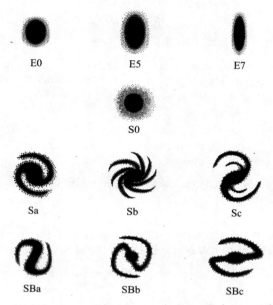

Fig. 10-4. Galaxy classification. The galaxies in the top row are ellipticals, the galaxies in rows 2 and 3 are spirals, and the galaxies in the bottom row are barred spirals.

Barred spiral. A barred spiral is very much like a spiral, except that the bulge is elongated, forming a bar across the center. Some people group barred spirals and spirals together in the same group. The barred spirals are also subdivided into three classes: SBa, SBb, and SBc. Once again, these subclasses indicate the relative size of the bulge.

2. ELLIPTICALS

Elliptical galaxies have no arms and no disk. They are big blobs of reddish, older stars. These are sub-classed as E0 . . . E3 . . . E7. E0 are the roundest type, whereas E7 are the most elliptical. Projection effects (which side of the galaxy you are seeing—the short end or the long side) may complicate the classification of an elliptical.

Ellipticals can also be sorted by size. Giant ellipticals are a few trillion parsecs across. Giant ellipticals contain trillions of stars. Dwarf ellipticals, on the other hand, are only about a thousand parsecs across, and contain millions of stars. There are many more dwarfs than giants, but the giants are so large that they actually contain most of the mass that's found in elliptical galaxies. Ellipticals contain very little gas and dust. Figure 10-4 shows a few elliptical galaxies.

CHAPTER 10 Galaxies and Clusters

The orbits of stars in elliptical galaxies are disordered, like the orbits of globular clusters in the halo of a spiral, or like the orbits of the stars in the bulge.

3. IRREGULARS

Irregular galaxies are all the galaxies that don't fit in the other categories. In general, they are thought to be the result of collisions between galaxies. They contain a lot of dust and gas, and consequently are full of young stars. There is no regular structure. The fraction of galaxies that are irregular is small.

DISTANCES TO GALAXIES

The distances to galaxies have been determined "step-wise," with the closest ones determined first, and then farther and farther ones. More distant galaxies have more uncertain distances, as a rule. In general, the distances to galaxies are determined by comparing objects in near galaxies with similar objects in more distant ones. For the most nearby galaxies, direct measurements of Cepheid variables yield good distances. Further out, the size of the largest HII region, or the average brightness of a globular cluster can be used. The Tully-Fisher relation is used to 100 million parsecs. This relation is a correlation between the luminosity of a galaxy and the breadth of the 21 cm line. Beyond this, the size of galaxies as measured on the sky and the brightest galaxy in a cluster are used to determine the distance. These methods are summarized in Table 10-1.

Table 10-1. Each distance-determination method is useful only for a certain range of distances.

Method	Distances to which the method is useful
Cepheid variables	10 million parsecs
Size of largest HII region, average brightness of globular cluster	25 million parsecs
Type I supernovae; Tully-Fisher relation (relates luminosity with breadth of 21 cm line)	100 million parsecs
Size of galaxies	100s of millions of parsecs
Brightest galaxy in a cluster	1,000 million parsecs

Beyond this, we must use **Hubble's Law**, an empirical relationship between the distance to a galaxy and the speed at which it is moving away from us. This relation is discussed in detail in the next chapter. Plots of the radial velocity of galaxies versus distance yield a straight line (Fig. 10-5). This line can be described by the equation

$$v = Hd$$

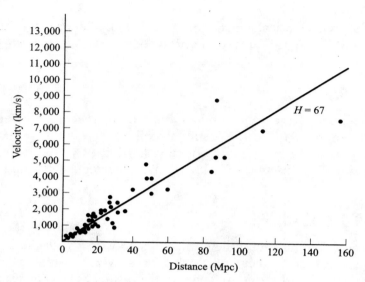

Fig. 10-5. Hubble's Law.

where v is the radial velocity, d is the distance to the galaxy, and H is the slope of the line, called Hubble's constant. Currently, the best estimate for H is 65 km/s/Mpc (1 Mpc = 1 Megaparsec = 1 million parsecs). Astronomers can find the distance to a galaxy simply by measuring its radial velocity, and using this law.

GALAXY DISTRIBUTION

Galaxies "cluster." For example, the Milky Way is part of a group of galaxies known as the Local Group. The Andromeda Galaxy is also a part of this group, and is our nearest large neighbor. There are about 20 galaxies in our Local Group, which are gravitationally bound to the Milky Way and the Andromeda Galaxy. Apart from the Milky Way and Andromeda, the rest of the Local Group is composed of mainly dwarf ellipticals, which are so small that their gravity does not strongly influence the Milky Way. The next nearest cluster is the Virgo cluster, which is 18 Mpc away, and 3 Mpc across. Virgo is a much larger cluster, and contains about 2,500 galaxies. There are thousands of galaxy clusters like Virgo. There are so many galaxies that if they are plotted on a map of the sky, it looks denser than the stars. Clusters also "cluster," and these clusters of clusters are called **superclusters**. In between these superclusters are great voids. The superclusters seem to come in sheets, so that the structure of the Universe is bubbly (although not with round bubbles), like a pile of soapsuds.

The origin of these walls and voids is poorly understood. It is not clear whether galaxies formed first, then gathered into clusters, or whether big cluster-sized clouds formed first, then galaxies condensed out of them.

CHAPTER 10 Galaxies and Clusters

Solved Problems

10.6. Suppose that we were to discover that all galaxies are actually twice as far away as we think. How would this affect Hubble's constant?

Hubble's Law says that the velocity and the distance are related by

$$v = Hd$$

If the distance increases by a factor of two, but the velocity remains the same, then that means that Hubble's constant must decrease by a factor of two. The currently accepted value (65 km/s/Mpc) would be revised to about 32 km/s/Mpc.

10.7. Astronomers recently claimed to have discovered the most distant galaxy known. How would they have determined this distance?

The distances to the farthest galaxies are determined using Hubble's Law. Astronomers would have measured the spectrum of the galaxy, and determined the radial velocity from the Doppler shifts of the emission lines. Then they would have used Hubble's Law to find the distance from this velocity, and the currently accepted value of Hubble's constant.

10.8. Suppose you measure the radial velocity of a galaxy to be 5,000 km/s. How far away is it?

To find the distance from the radial velocity, use Hubble's Law. The current best estimate of H is 65 km/s/Mpc.

$$v = Hd$$
$$d = \frac{v}{H}$$
$$d = \frac{5{,}000 \text{ km/s}}{65 \text{ km/s/Mpc}}$$
$$d = 80 \text{ Mpc}$$

The galaxy is 80 Mpc, or eighty million parsecs away.

10.9. Suppose a galaxy is 150 Mpc away. What is its radial velocity?

To find the velocity from the distance, use Hubble's Law, with the current best estimate of H, 65 km/s/Mpc.

$$v = Hd$$
$$v = 65 \text{ km/s/Mpc} \cdot 150 \text{ Mpc}$$
$$v = 9{,}800 \text{ km/s}$$

The galaxy is observed to be moving away at nearly 10,000 km/s.

10.10. At the same distance from the center, the rotation speed of one galaxy is twice that of another. What is the ratio of their masses?

Use the circular velocity equation, and form a ratio of the equations for each galaxy. Since the radii are equal, they will cancel.

$$\frac{v_1}{v_2} = \frac{\sqrt{\frac{GM_1}{R_1}}}{\sqrt{\frac{GM_2}{R_2}}}$$

$$\frac{v_1}{v_2} = \frac{\sqrt{\frac{GM_1}{R_1}}}{\sqrt{\frac{GM_2}{R_2}}}$$

$$\frac{v_1}{v_2} = \sqrt{\frac{M_1}{M_2}}$$

$$\frac{M_1}{M_2} = \left(\frac{v_1}{v_2}\right)^2$$

$$\frac{M_1}{M_2} = \left(\frac{v_1}{2v_1}\right)^2$$

$$\frac{M_1}{M_2} = \frac{1}{4}$$

The mass of the galaxy with the slower rotation speed is 1/4 the mass of the galaxy with the faster rotation speed.

10.11. Why can't Cepheid variables be used to find the distances to galaxies 100 Mpc away?

To use a Cepheid variable to determine the distance to a galaxy, it must be bright enough to observe. At 100 Mpc, Cepheid variables are too faint to distinguish their light from the rest of the light from the galaxy.

10.12. The Andromeda Galaxy is approaching the Milky Way at 266 km/s. The galaxies are approximately 1 million pc apart. How long will it be until they collide? (Assume the speed of approach remains the same.)

First, convert 1 million pc to kilometers by multiplying by 3×10^{19} km/pc, to find a distance of 3×10^{19} km. To find the time it will take the Andromeda Galaxy to travel that distance at 266 km/s, divide the distance by the speed:

$$t = \frac{d}{v}$$

$$t = \frac{3 \times 10^{19} \text{ km}}{266 \text{ km/s}}$$

$$t = 1.1 \times 10^{17} \text{ s}$$

$$t = 3.5 \times 10^9 \text{ yr}$$

The Andromeda Galaxy will not collide with the Milky Way for 3.5 billion years. The collision will actually occur somewhat sooner than this, as the two galaxies may be expected to speed up as they approach each other.

10.13. A Cepheid variable star in the Virgo cluster has an absolute magnitude of −5, and an apparent magnitude of 26.3. How far away is the Virgo cluster?

CHAPTER 10 Galaxies and Clusters

Use the magnitude–distance relation from Chapter 7.

$$m = M + 5\log\left(\frac{d(\text{pc})}{10}\right)$$
$$d(\text{pc}) = 10^{(m-M+5)/5}$$
$$d(\text{pc}) = 10^{(26.3-(-5)+5)/5}$$
$$d(\text{pc}) = 10^{36.3/5}$$
$$d = 1.8 \times 10^7 \text{pc}$$

The distance to the Virgo cluster is 18×10^6 pc (18 million pc, or 18 Mpc).

10.14. Two galaxies orbit each other with an orbital period of 50 billion years. The distance between them is 0.5 million pc. What is the mass of the pair?

First, convert 50 billion years into seconds, to get a period of 1.58×10^{18} seconds. Then, convert 5×10^5 parsecs to 1.5×10^{19} kilometers. Use Kepler's Third Law, from Chapter 1:

$$P^2 = \frac{4\pi^2 a^3}{G(m+M)}$$
$$(m+M) = \frac{4\pi^2 a^3}{GP^2}$$
$$(m+M) = \frac{4\pi^2 (1.5 \times 10^{22} \text{ m})^3}{6.67 \times 10^{-11} \text{ m}^3/\text{kg/s}^2 (1.58 \times 10^{18} \text{ s})^2}$$
$$(m+M) = \frac{1.33 \times 10^{68} \text{ m}^3}{1.67 \times 10^{26} \text{ m}^3/\text{kg}}$$
$$(m+M) = 8 \times 10^{41} \text{ kg}$$

The combined mass of the two galaxies is 8×10^{41} kg, or about 4×10^{11} solar masses.

10.15. Using Hubble's Law, calculate how long it will take for the distance between the Milky Way and the Coma cluster ($d = 113$ Mpc) to double.

The velocity of the Coma cluster is given by Hubble's Law to be $v = Hd$. The time it will take for the Coma cluster to travel the same distance, moving at that velocity, is

$$t = \frac{d}{v}$$
$$t = \frac{d}{Hd}$$
$$t = \frac{1}{H}$$

This time is known as the Hubble Time (more about that in Chapter 11), and is equal to

$$t = \frac{1}{H}$$
$$t = \frac{1}{65 \text{ km/s/Mpc}}$$

Converting Mpc to km, by dividing the denominator by 3×10^{19} km/MPc, gives $t = 4.6 \times 10^{17}$ s, or 1.5×10^{10} (15 billion) years.

Active Galaxies and Quasars

An active galaxy is a galaxy with a nucleus that produces an exceptionally large amount of energy, up to 10^{15} solar luminosities. This is about 10,000 times the amount of energy produced by the entire Milky Way Galaxy. The spectrum of an active galaxy is quite flat compared with the spectrum of a star, and in many cases contains **both** broad and narrow emission lines. The luminosity in the radio waves through far-infrared is dominated by synchrotron radiation—radiation produced by electrons spiraling around a magnetic field line. The luminosity in the ultraviolet is dominated by the "blue bump" produced by hot (10,000 K), dense gas. And the near-infrared radiation is dominated by emission from cool dust (see Fig. 10-6).

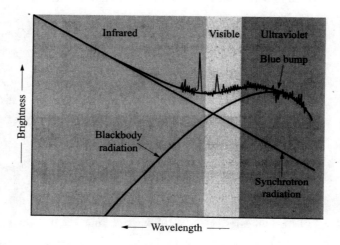

Fig. 10-6. The spectrum of an active galaxy.

The nuclei of active galaxies are only a few parsecs in size. This is determined from the fluctuation time, τ, of the total luminosity. If the luminosity fluctuates rapidly, the object must be small, so that the entire object gets the message to start varying at about the same time, and the light from the more distant parts does not have to travel significantly farther (therefore take longer) to reach the observer. If the luminosity fluctuates slowly, the object is large, because it took longer for the signal to get out to the more distant side, and for the response of that distant side to travel back across the intervening space (see Fig. 10-7). An estimate of the size of the object, D, is given by the product of the fluctuation time, τ, times the speed of light, c:

$$D = c\tau$$

CHAPTER 10 Galaxies and Clusters

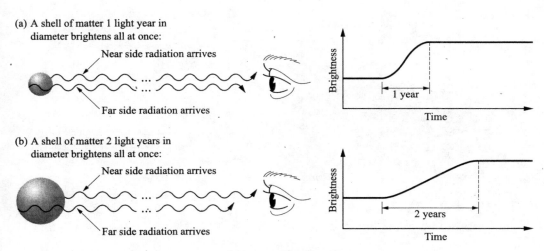

Fig. 10-7. The size of an object may be determined from the speed of the variations.

Thus, a quasar fluctuating every 100 days will have a size of

$$D = c\tau$$
$$D = 3 \times 10^8 \cdot 10^2 \cdot 864{,}000 \text{ s/day}$$
$$D = 2.6 \times 10^{16} \text{ m}$$
$$D = 1 \text{ pc}$$

There are several different varieties of active galaxies, depending on whether they show broad and/or narrow emission lines, whether the host galaxy can be seen, whether they vary rapidly, etc. Three main categories are radio galaxies, Seyfert galaxies, and quasars.

It is uncertain whether these types represent truly distinct objects, or the same objects viewed at varying distances or orientations (Fig. 10-8). In radio galaxies, the radio waves come primarily from two lobes protruding on opposite sides of a central elliptical galaxy. Seyfert galaxies eject mass, like radio galaxies, but have a spiral galaxy at the central portion. The nucleus of the spiral galaxy is much more luminous than nuclei of ordinary spirals. The radiation from the nucleus fluctuates very rapidly (τ is on the order of 1 hour, so the size of the nucleus is about 7 AU). Quasars (quasi-stellar radio sources) are similar to the other active galaxies and their intensities fluctuate with τ ranging from months to hours, indicating small sizes. The red shift in some spectral lines of quasars are higher than any object, indicating that quasars have the highest recession speed. From Hubble's Law, we know that the recession speed increases with distance; hence, we conclude that quasars are the most distant objects.

The staggering luminosities of these objects may be explained by the presence of a supermassive black hole in the center. As stars, gas, and dust fall in towards the black hole, conservation of angular momentum forces them into an accretion disk, where they spiral around the black hole. As they turn, they emit ultraviolet and X-ray radiation, which is subsequently absorbed by the surrounding dust and gas,

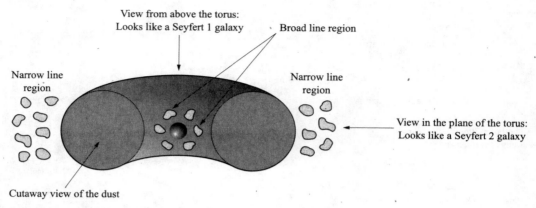

Fig. 10-8. Dusty torus model of Seyfert galaxies.

heating it, so that the radiation is re-emitted as continuum radiation at lower temperatures.

The radiation produced by the infalling material pushes outward on the outer rim of the accretion disk. For a given mass, the luminosity has an upper limit, beyond which all the outer matter would be swept away by the radiation. This is called the **Eddington Luminosity**, L_{Edd}, and is given by

$$L_{Edd} = 30,000 \frac{M}{M_{Sun}} L_{Sun}$$

where M is the mass of the black hole, M_{Sun} is the mass of the Sun, and L_{Sun} is the luminosity of the Sun. This equation can be turned around to give the **lowest** possible mass that a central black hole could have to produce the luminosity of an active galaxy. The mass could be higher, but not lower than this value, and still produce this luminosity.

Active galaxies often have jets (Fig. 10-9). These jets are produced when blobs of ionized matter are carried away from the accretion disk along the magnetic field lines. The spiraling material produces a magnetic field of its own which compresses the outflowing material into a thin jet.

Quasars were much more common in the distant past than they are today, with nearly all quasars found farther than $2^{-1/3}$ (about 80%) the distance to the most distant one. (If they were equally common today, we would expect to find half of them nearer than this distance and half farther. The $-1/3$ power comes in because it's really the *volumes* that must be compared, not simply the distances.) The luminosity of quasars appears to have declined over time. Perhaps the more nearby Seyferts are simply modern versions of more distant quasars. The variation of quasars with distance (therefore time) is one of the best arguments that the Universe has evolved, and is different today than at some distant time in the past. However, in the **very** early Universe, no quasars are observed. The implication is that it took some time for the host galaxies and the supermassive black holes to form and start producing the large luminosities by which we identify quasars.

In addition to being interesting objects in and of themselves, observations of active galaxies yield information about the intergalactic medium and foreground

CHAPTER 10 Galaxies and Clusters

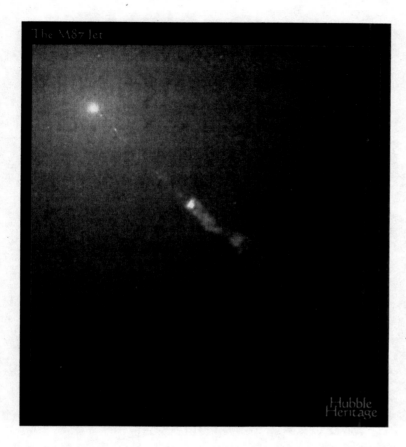

Fig. 10-9. AGN jets in M87. (Courtesy of the Hubble Heritage Project, STScI.)

galaxies. Absorption lines of hydrogen indicate the presence of hydrogen in the intervening medium. Most of this hydrogen is gathered into clouds, which were more common in the early Universe than today, probably because these clouds have been integrated into galaxies.

Solved Problems

10.16. Figure 10-10 shows plots of the luminosities of several active galaxies versus time. Which of these active galaxies is smallest? Which is largest?

The galaxy with the fastest variation is the smallest galaxy, so that is galaxy C. Galaxy D has the slowest variation, and so is the largest galaxy.

CHAPTER 10 Galaxies and Clusters

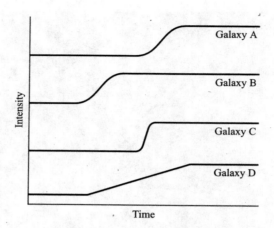

Fig. 10-10. Variation of active galaxies versus time.

10.17. What is the Eddington Luminosity of a quasar with a black hole mass of 500,000,000 M_{Sun}?

Use the Eddington Luminosity equation. Since we will divide by M_{Sun} in this equation, we do not need to convert the mass to kilograms first.

$$L_{Edd} = 30,000 \left(\frac{M}{M_{Sun}}\right) L_{Sun}$$

$$L_{Edd} = 30,000 \left(\frac{5 \times 10^8 M_{Sun}}{M_{Sun}}\right) L_{Sun}$$

$$L_{Edd} = 1.5 \times 10^{13} L_{Sun}$$

The Eddington Luminosity of a quasar with a black hole mass of $5 \times 10^8 \, M_{Sun}$ is 1.5×10^{13} times the luminosity of the Sun.

10.18. What is the smallest possible mass of a quasar with luminosity $10^{15} \, L_{Sun}$?

Turn the Eddington Luminosity equation around to solve for M, and substitute:

$$M = \frac{L_{Edd} \cdot M_{Sun}}{30,000 \cdot L_{Sun}}$$

$$M = \frac{10^{15} L_{Sun} \cdot M_{Sun}}{30,000 \cdot L_{Sun}}$$

$$M = 3.3 \times 10^{10} M_{Sun}$$

The minimum mass of the quasar is 33 billion solar masses.

10.19. Material spiraling into black holes loses about 30% of the mass as light that escapes the system. How much energy is emitted when an object the mass of the Earth spirals into the black hole?

First, calculate the total mass converted to energy.

$$M = 0.30 \cdot M_{Earth}$$

$$M = 1.8 \times 10^{24} \text{ kg}$$

CHAPTER 10 Galaxies and Clusters

Now, find the energy using $E = mc^2$:

$$E = mc^2$$
$$E = 1.8 \times 10^{24} \cdot (3 \times 10^8)^2 \text{ J}$$
$$E = 1.6 \times 10^{41} \text{ J}$$

There are 1.6×10^{41} J released when an Earth-sized object falls into a black hole.

Supplementary Problems

10.20. Gas near the center of the Milky Way ($R = 0.1$ pc) has been observed to have an orbital speed of 700 km/s. What is the mass of the material inside this radius?

Ans. $1.1 \times 10^7 \, M_{Sun}$

10.21. How much energy is released when an object of 1 M_{Sun} falls into a black hole?

Ans. 5.4×10^{46} J

10.22. What is the orbital speed of a star 10 pc from the center of a $1 \times 10^9 \, M_{Sun}$ galaxy, if its orbit is circular?

Ans. 660 km/s

10.23. What is the orbital period of the star in Problem 10.21?

Ans. 93,000 years

10.24. Suppose that you observe a new kind of galaxy that has a rotation curve that falls off with distance from the center. How is this galaxy different from all other known galaxies?

Ans. It has no dark matter

10.25. You read an article claiming that astronomers have discovered the distances to five dwarf galaxies in the Local Group. What distance determination method was most likely used?

Ans. Cepheid variables

10.26. You observe a galaxy with a redshift ($\Delta\lambda/\lambda$) equal to 2. What is the apparent velocity of this galaxy?

Ans. 6×10^8 m/s

10.27. Is the galaxy in Problem 10.26 approaching or receding? How far away is it?

Ans. Receding, 0.9 Mpc

10.28. What is the Eddington Luminosity of a quasar of mass $1 \times 10^9 \, M_{Sun}$?

Ans. $3 \times 10^{13} \, L_{Sun}$

10.29. Suppose you determine that a quasar has luminosity 10^{20} times that of the Sun. What is minimum mass of this quasar?

Ans. $3 \times 10^{15} \, M_{Sun}$

10.30. A quasar's luminosity varies with $\tau = 2$ years. How large is this quasar?

Ans. $6 \times 10^{15} \, \text{m} = 0.2 \, \text{pc}$

Cosmology

Cosmology is the study of the history of the Universe as a whole, both its structure and evolution. The study assumes that over large distances the Universe looks essentially the same from any location (the Universe is homogeneous) and that the Universe looks essentially the same in all directions (the Universe is isotropic). The assumptions are known as the **cosmological principle**. In addition, it is assumed that the same laws of physics hold everywhere in the Universe.

Reliable astronomical records are available for the last 100 years or so. This time interval is minute by astronomical scales, which are typically on the order of 10 billion years. To study the evolution of the Universe, astronomers rely on observations of distant objects. For example, the Andromeda Galaxy (M31) is about 2 million light years away. This means that our photographs of M31 show the galaxy as it was 2 million years ago. The Virgo cluster of galaxies is about 50 million light years away, and the information presently received by our telescopes describe the state of the cluster as it was 50 million years ago. Information received from galaxies that are billions of light years away pertains to the state of these objects as they were billions of years ago, when they were much younger, possibly just forming. Receiving information from distant objects is equivalent to receiving information from the past. This information is vital in reconstructing the evolutionary stages of the galaxies, and the Universe as a whole. As the development of testable hypotheses is limited by our present instrumental capabilities, the theories proposed are speculative.

Hubble's Law

Edwin Hubble determined that there is a relation between the distance of various galaxies and their radial velocity, derived from the Doppler shift of absorption lines in the spectrum of the galaxies. The spectral lines appear red-shifted for all galaxies. This result is very significant: all galaxies are moving away from us, i.e., they are receding. The amount of red shift in the absorption is proportional to the distance of the galaxy. The relation, known as Hubble's Law, is given by

$$v = Hd$$

where v is the radial recession velocity, d is the distance, and H is the Hubble constant. Current best estimates of H are around 65 km/s/Mpc. Note that the units of H are actually inverse time. The only galaxies that deviate from Hubble's Law are those so close that gravity dominates their behavior.

Hubble's Law and the Expansion of the Universe

If the galaxies are just moving randomly through space, some near ones would move rapidly and some distant ones would move slowly. About half as many would approach as recede. The fact that all the galaxies are moving away from us does not mean that we are in a special place in the Universe. Hubble's Law could be measured at any location in the Universe, with the same result. This means that space itself is expanding, because the more distant the galaxies are, the faster they are moving away. A two-dimensional analog is the surface of an expanding balloon. If you blow up a balloon part way, and mark little dots on it to indicate galaxies, then continue to blow it up, you will find that the distance between the dots that were most separated increased more than the distance between dots that were close together. In the same way, more distant galaxies move away faster, and so we know that the space in between grew. Fortunately, there is more evidence than simply the Hubble Law, and this evidence has convinced astronomers that the Universe is expanding. This evidence will be discussed later in this chapter.

Hubble's Law and the Age of the Universe

According to Hubble's Law, the recession velocity v divided by the distance d is equal to the constant $1/H$, the same for all galaxies. The ratio v/d gives the time it took for the Universe to expand to the present state. This implies that $1/H$ is the age of the Universe. Using $H = 65$ km/s/Mpc, and values from Appendix 2, we can estimate the age of the Universe to be about 15 billion years.

An important assumption in this estimate is that the rate of expansion is constant. The gravitational interaction between galaxies would tend to slow down the expansion. On the other hand, observations of Type Ia supernovae in very distant galaxies indicate recession speeds lower than predicted by Hubble's Law. But for very distant galaxies, the measured speed corresponds to their speeds in the distant past when the observed spectrum was emitted, i.e., the speeds in the past were slower, so the expansion has accelerated. The expansion age of the Universe is often referred to as the Hubble Time, and depends on the value of Hubble's constant, which may include significant uncertainty.

CHAPTER 11 Cosmology

The uncertainties lead to some paradoxical results. For example, some globular clusters and white dwarfs seem to be older than 15 billion years, which is inconsistent with the Hubble Time. The error may result either from stellar evolution calculations, or in the determinations of distances and velocities, or in the assumptions. The Universe is probably not younger than 15 billion years old. Though it may be older, it's probably not very much older.

Hubble's Law and the Size of the Universe

Using the Hubble Time we can use conversions from Appendix 2 to estimate the size of the Universe as

$$R = ct = 15 \text{ billion light years}$$

The Big Bang

One implication of Hubble's Law is that the entire Universe was once much denser than it is now, so that all the mass was much closer together. But Hubble's Law is not the only piece of evidence.

1. **The Universe contains mass, and gravity is the only force acting on the large scale of distances, and tends to pull mass together.** This observation implies that at least one of three things is true:
    (i) The Universe is of finite age. Gravity hasn't had enough time yet to collapse the Universe.
    (ii) The Universe is infinite in extent. There's no center for all of the mass to fall to.
    (iii) The Universe is expanding faster than its own escape velocity. All the galaxies have enough speed to overcome their mutual gravitational attraction.
2. **The night sky is dark.** This is also known as **Olbers' paradox**. Given an infinite Universe, the night sky ought to be as bright as the day. This is analogous to standing in the forest and looking through the trees. If the forest is large enough, then you observe nothing through the trees but more trees. This implies that at least one of four things is true:
    (i) The Universe is of finite age. Light hasn't had enough time yet to get here from the most distant places.
    (ii) The Universe is of finite extent. There aren't actually enough objects out there to fill up the sky.

(iii) The Universe is expanding/contracting so fast that the light is getting Doppler shifted out of any band of the electromagnetic spectrum we can observe.

(iv) Stars are a "new" phenomenon. The Universe only recently started to produce light.

3. **The Universe is expanding.** This implies that
    (i) The Universe is of finite age.
    (ii) Because we can measure this, we know that 1(iii) and 2(iii) above are incorrect.

4. **The cosmic microwave background radiation (CMBR) exists.** It is a perfect blackbody spectrum at 2.74 K. The CMBR is the same (isotropic) to a fairly high degree of precision in every direction in the sky, with only very small deviations (anisotropies). The existence of this background implies that at one time the whole Universe was the same temperature.

5. **The helium abundance.** The amount of helium in the Universe is about 25% of the total amount of matter. By stellar nucleosynthesis alone, it should be only about 10%. This implies that, at some time, there was a massive burning of H into He.

In summary, the Universe is finite in age. It may or may not be finite in extent. At one time, the Universe was hot, compressed, and all the same temperature. Thus, the Big Bang: a hot beginning.

The Big Bang was the beginning of time and space. Asking what happened before the Big Bang is a lot like asking "Where is north from the North Pole?" Anywhere from the North Pole is south. Any when from the Big Bang is later in time. The Big Bang is the beginning of time. There is no "before." Where did the Big Bang happen is a similarly impossible question. Space itself was created in the Big Bang. There was no where "before" that. In a sense, the Big Bang happened everywhere. This is why the CMBR is the same everywhere. All the points in the Universe were the same point, and all of them were the same temperature.

The Universe does not have a "center," whether or not it is infinite. A two-dimensional analog to the finite Universe is the surface of a balloon: all points on the surface are equivalent. There are no edges, and there is no center. Note also that this means that there is no "outside."

1. There is no center. The Universe is not expanding "away" from anything.
2. There is no "before." Time began at the Big Bang.
3. The Universe is not expanding "into" anything. Space is created by the expansion of the Universe.

THE RADIATION-DOMINATED UNIVERSE

Astronomers have a good understanding of the beginning of the Universe, all the way back to about 10^{-34} seconds after it began. The grand unified theories (GUT) that describe the ways that the known forces of physics operated in the very early

CHAPTER 11 Cosmology

Universe are not experimentally verifiable by present means. The radiation-dominated era lasted from about 10^{-34} seconds to about 10^5 years. The temperature of the Universe cooled during this time from about 10^{32} K to 10^3 K. The energy of the Universe during this era was mainly in the form of photons; hence the name radiation era. Photons of high energy (gamma rays) can create pairs of particles of matter–antimatter, e.g., electron–positron pairs. The energy of the gamma ray required to produce a particle–antiparticle pair can be calculated from Einstein's mass–energy formula: $E = mc^2$, where m is the mass of the particle (also the antiparticle) and c is the speed of light. We can use this equation to estimate the energy and the temperature required for the creation of particle–antiparticle pairs. For example, the gamma ray required to produce an electron–positron pair should be at least equal to $2(m_e c^2)$ plus the kinetic energy of the particles (m_e is the mass of the electron). The factor 2 accounts for the fact that two particles are created. Neglecting the kinetic energy of the particles, and using values from Appendix 2, we can estimate the energy required to be $E = 1.6 \times 10^{-13}$ joules. From Wien's Law (Chapter 1), a blackbody (the Universe) at temperature T has a wavelength maximum of λ_{max} (m) $= 0.0029/T$. Using this value in Planck's relation for the energy of a gamma-ray photon, we find

$$E = hf = \frac{hc}{\lambda}$$

$$E = \frac{hcT}{0.0029}$$

Solving for the temperature, T, we have

$$T = 0.0029 \frac{E}{hc}$$

$$T = 0.0029 \frac{1.6 \times 10^{-13}}{(6.6 \times 10^{-34} \cdot 3 \times 10^8)} \text{ K}$$

$$T = 2.3 \times 10^9 \text{ K}$$

A similar estimate for protons–antiprotons gives $T = 5 \times 10^{12}$ K.

10^{-6} seconds after the Big Bang. The Universe cooled enough so that the gamma rays no longer had enough energy to make protons, and a few seconds later, the temperature was too low for the creation of electron/positron pairs. Pair annihilation continued. The density of the Universe decreased enough that neutrinos stopped interacting with matter, and propagated almost freely.

1 to > 300 seconds after the Big Bang. The protons and neutrons fused to form deuterium nuclei (heavy hydrogen). The photons still had enough energy to break deuterium apart. At about 100 seconds, the temperature had dropped to 1 billion K, and the deuterons began to survive, since a typical photon no longer had enough energy to tear them apart. Deuterium nuclei fused to form helium (see the proton–proton chain, Fig. 8-1). In addition, minute amounts of lithium and beryllium were formed. The observed abundances of these elements in primordial matter is one of the most important pieces of evidence in support of the Big Bang.

THE MATTER-DOMINATED UNIVERSE

Until 100,000 years after the Big Bang, the amount of energy in the form of radiation far exceeded the amount of energy in the form of mass. Once the temperature falls to a few thousand K, however, two things happen: matter begins to dominate, and the Universe becomes transparent to radiation. Until this time, the Universe was opaque. Astronomers can never observe the first 100,000 years, no matter what technical improvements in telescopes occur. This is the "current" epoch of the Universe. Once the temperature was low enough for atoms to survive, the Universe became transparent. The time when this happened is called the "decoupling epoch" (because the radiation and the matter become mostly decoupled from each other), or the "recombination epoch" (because the electrons and the nuclei "recombine"). Nearly all of the radiation that existed at this time is still traveling through the Universe. This is what we see when we observe the cosmic background radiation—very strongly red-shifted light, which has been traveling through the Universe since about 100,000 years after the Big Bang, and is just now reaching us. Galaxies and stars began to form at about 1 billion years after the Big Bang.

THE CURVATURE OF SPACE-TIME

Space is bent, not just locally by gravity, but possibly in a larger sense, as the entire Universe may have a curvature. There are three possibilities for the curvature of the Universe. Visualizing these curvatures is far easier in two dimensions than in four.

1. **0 curvature (flat).** In two dimensions, this can be visualized as a piece of paper. A flat Universe is infinite. Someone traveling in one direction, just keeps going that way forever and ever, and never comes back around to where they started. But still, there are no boundaries, and therefore no center. Angles in triangles add up to 180°. The circumference of a circle is always $2\pi R$. All of the Euclidean geometry rules apply.
2. **Positively curved.** In two dimensions, this can be visualized as the surface of a sphere. In this case, the Universe is finite, yet unbounded. Someone traveling in one direction long enough comes back to where they started. In positively curved space, there are no straight lines. The shortest distance between two points is along a great circle. The circumference of circles is $< 2\pi R$. The sum of the angles in a triangle adds up to more than 180°.
3. **Negatively curved.** In two dimensions, this is often called saddle shaped. Like flat space, negatively curved space is infinite in extent, and has no boundaries. In this kind of space, the angles of a triangle add up to **less** than 180° and the circumference of a circle is $> 2\pi R$.

General relativity says that space is curved by the presence of matter. Therefore, the density determines the curvature. If the actual density of the Universe, ρ, is larger than a critical density, ρ_c, then the Universe will have positive curvature. If

CHAPTER 11 Cosmology

$\rho < \rho_c$, then the Universe will have negative curvature. If $\rho = \rho_c$, then the Universe will be flat. The critical density is extremely small ($\sim 10^{-26}$ kg/m^3). The most recent results indicate that the density of the Universe is pretty close to flat, and ρ/ρ_c is between 0.1 and 2. That is, the density is close to the critical density. ρ/ρ_c is also known as Ω_0 (pronounced "omega-naught").

The future of the Universe depends on the density, and therefore on the curvature. If astronomers can determine either the density or the curvature, then they can predict how the Universe will end. Hubble's constant determines the expansion rate of the Universe, and the average density describes the self-gravitation and curvature. If H is small, and the density is large, the Universe will recollapse. If H is large, and the density is small, the Universe will keep expanding indefinitely. The dividing line between the two occurs when $\Omega_0 = 1$, where the Universe expands until it reaches equilibrium, then stops.

THE BIG CRUNCH

If $\Omega_0 > 1$, then the Universe will slow down, stop expanding, and begin to contract. As this occurs, the nearby galaxies will become blue-shifted. The more distant galaxies will still appear as they were in the past, before contraction began. However, as time passes, more and more distant galaxies will begin to be blue-shifted, until we are observing the Hubble Law in reverse.

Gravity accelerates the collapse, and the density and the temperature of the Universe rise. Eventually, the density and temperature get so high that the Universe can "re-explode," forming a new Big Bang.

THE BIG FREEZE

If $\Omega_0 < 1$, the Universe will expand indefinitely. Binary stars and galaxy clusters collapse into each other, and the stars gradually burn out, and form black holes. On time scales of 10^{32} years, protons in the Universe will decay into radiation. All particles and black holes will "evaporate" over time. Eventually, all the mass in the Universe disappears. All that is left is radiation, which reddens and weakens forever.

INFLATION

While the Big Bang model is successful in explaining many things about the Universe, it fails to explain three major observations:

1. The extreme uniformity of the CMBR in different regions of space, which were widely separated even when the Universe first became transparent. If these regions were widely separated, how did they come to the same temperature? This is the horizon problem.
2. The value of Ω_0 should remain either greater than or less than 1. The $\Omega_0 = 1$ case is sensitive to small perturbations, and is unstable. Any deviation

from flatness should have grown to noticeable extremes by now. This is the flatness problem.

3. The existence of large-scale structure, like walls of clusters of galaxies and voids, which are unexplained in the basic Big Bang theory. This is the structure problem.

Inflation is an adjustment to the basic Big Bang which has been proposed in order to solve these problems. Essentially, inflation removes the assumption that the Universe has always expanded at the same rate. Inflation says that between about 10^{-34} seconds and 10^{-32} seconds after the Big Bang, the Universe underwent an extremely accelerated expansion, caused by the separation of various fundamental forces into individual forces. Prior to this time, these three forces (called electromagnetic, strong, and weak) were unified at the higher temperatures and pressures, and indistinguishable from one another, because they have precisely the same strengths. This rapid period of expansion solves the three problems remaining in the Big Bang.

Adding inflation to the Big Bang theory means that the Universe prior to 10^{-34} seconds was much denser than previously thought. This solves the horizon problem. With inflation, the regions of space that emitted the CMBR were actually quite close together when the Universe became transparent. Therefore it is reasonable that they might have been the same temperature, since they could exchange heat quite easily.

The flatness problem is solved by inflation because during that rapid expansion, the Universe expanded faster than the speed of light, so a large fraction of the Universe now exists beyond what will ever be observable. Consider yourself on the Earth. You can see about 1 mile all around you, and the Earth looks pretty flat on that scale. However, if the Earth were smaller, you would see significant curvature over the same 1 mile scale. In the same way, because our view of the Universe is limited, we see that it is locally quite flat, even if on larger scales (or when the whole Universe was smaller), it had noticeable positive or negative curvature.

Minute deviations from uniformity that existed prior to inflation were magnified during inflation. The process occurred so quickly that the non-uniformities would not have had time to become "smoothed out," and imprinted themselves on the large-scale structure of the Universe.

Solved Problems

11.1. Suppose that the Hubble constant is found to be 70 km/s/Mpc. What is the Hubble time for this value of the Hubble constant?

CHAPTER 11 Cosmology

The Hubble Time is given by $1/H$. First, let's convert H into units of time. (Recall that $1\text{ pc} = 3.09 \times 10^{13}\text{ km}$.)

$$H = 70\text{ km/s/Mpc}$$
$$H = 70\text{ km/s/Mpc} \cdot (1\text{ Mpc}/10^6\text{ pc}) = 70 \times 10^{-6}\text{ km/s/pc}$$
$$H = 70 \times 10^{-6}\text{ km/s/pc} \cdot (1\text{ pc}/3.09 \times 10^{13}\text{ km})$$
$$H = 22.65 \times 10^{-19}\text{ s}^{-1}$$

Now, invert to find the Hubble Time.

$$\frac{1}{H} = \frac{1}{70\text{ km/s/Mpc}}$$
$$\frac{1}{H} = \frac{1}{22.65 \times 10^{-19}\text{ s}^{-1}}$$
$$\frac{1}{H} = \frac{1}{22.65} \times 10^{19}\text{ s}$$
$$\frac{1}{H} = 0.04414 \times 10^{19}\text{ s}$$
$$\frac{1}{H} = 4.414 \times 10^{17}\text{ s}$$

Convert the seconds to years, and truncate to the correct number of significant figures (recall that $1\text{ yr} = 3.16 \times 10^7\text{ s}$).

$$\frac{1}{H} = 4.414 \times 10^{17}\text{ s} \cdot (1\text{ yr}/3.16 \times 10^7\text{ s})$$
$$\frac{1}{H} = 1.397 \times 10^{10}\text{ yr}$$
$$\frac{1}{H} = 13.97 \times 10^9\text{ yr} = 13.97\text{ billion years}$$
$$\frac{1}{H} = 14\text{ billion years}$$

A value of 70 km/s/Mpc gives a Hubble Time of 14 billion years.

11.2. Using the Hubble Law, we can derive an age of the Universe of about 15 billion years. But this assumes the Universe is empty, so that gravity is not slowing the expansion. If the Universe is flat, and the density is the same as the critical density, the age of the Universe is about two-thirds of the Hubble Time. What is the value of the Hubble constant in a flat, critical density Universe?

$$\text{Age} = 2/3 \cdot \frac{1}{H}$$
$$H = 2/3 \cdot \frac{1}{\text{Age}}$$
$$H = 2/3 \cdot \frac{1}{15 \times 10^9\text{ yr}}$$
$$H = \frac{2 \times 10^{-9}}{3 \cdot 15\text{ yr}}$$
$$H = 0.0444 \times 10^{-9}\text{ yr}^{-1}$$

Fix the units:

$$H = 0.0444 \times 10^{-9} \, \text{yr}^{-1}$$
$$H = 0.0444 \times 10^{-9} \, \text{yr}^{-1} \cdot (1 \, \text{yr}/3.16 \times 10^7 \, \text{s})$$
$$H = 0.0141 \times 10^{-16} \, \text{s}^{-1}$$
$$H = 0.0141 \times 10^{-16} \, \text{s}^{-1} \cdot (3.09 \times 10^{19} \, \text{km/Mpc})$$
$$H = 0.0434 \times 10^3 \, \text{km/s/Mpc}$$
$$H = 43.4 \, \text{km/s/Mpc}$$

For a 15-billion-year-old flat, critical density Universe, the Hubble constant has a value of 43 km/s/Mpc. This is at the low end of current estimates of the Hubble constant.

11.3. Suppose we observe a galaxy 500 Mpc away, which is moving away from us with a velocity of 30,000 km/s. If the speed has been constant throughout time, when did the Big Bang happen?

The key point is that when the Big Bang happened, the galaxy would have been at 0 distance from the Milky Way. We begin with the basic equation:

$$t = \frac{d}{v}$$
$$t = \frac{500 \, \text{Mpc}}{30{,}000 \, \text{km/s}}$$
$$t = 0.01667 \, \text{Mpc} \cdot \text{s/km}$$

Now fix the units:

$$t = 0.01667 \, \text{Mpc} \cdot \text{s/km} \cdot (3.09 \times 10^{19} \, \text{km/Mpc})$$
$$t = 0.0515 \times 10^{19} \, \text{s}$$
$$t = 0.0515 \times 10^{19} \, \text{s} \cdot (1 \, \text{yr}/3.16 \times 10^7 \, \text{s})$$
$$t = 0.0163 \times 10^{12} \, \text{yr}$$
$$t = 16 \times 10^9 \, \text{yr} = 16 \, \text{billion years}$$

If a galaxy like this were observed, it would imply that the Big Bang happened approximately 16 billion years ago. This is comparable to current estimates of the age of the Universe, so it would not be unusual to observe a galaxy at this distance with this recession velocity.

11.4. If $H = 65 \, \text{km/s/Mpc}$, the critical density is $8 \times 10^{-27} \, \text{kg/m}^3$. How much mass would be enclosed in a sphere with the radius of the Earth's orbit ($R = 1 \, \text{AU}$)? Assume a flat Universe.

$$V = \frac{4}{3} \cdot \pi \cdot R^3$$
$$V = \frac{4}{3} \cdot \pi \cdot (1.5 \times 10^{11} \, \text{m})^3$$
$$V = 14.137 \times 10^{33} \, \text{m}^3$$

Multiply the density by the volume to get the mass enclosed ($\rho = \rho_c$ in a flat Universe):

$$m = \rho \cdot V$$
$$m = (8 \times 10^{-27} \, \text{kg/m}^3) \cdot (14.137 \times 10^{33} \, \text{m}^3)$$
$$m = 113 \times 10^6 \, \text{kg}$$

CHAPTER 11 Cosmology

The mass enclosed in a sphere of radius 1 AU is 100 million kilograms. This mass is quite small when compared to the mass of the Earth ($M_E \sim 6 \times 10^{24}$ kg). This means that the Universe has a **really** low density.

11.5. Compare flat universes with positively curved universes. Are they closed? finite? do they have centers?

A flat universe is infinite, and open. Because it is infinite, it doesn't make sense to ask whether it has a center.
Conversely, a positively curved universe is finite and closed. But it also has no center, and is unbounded (has no edges).

11.6. If the density of the Universe is less than critical ($\rho < \rho_c$), what is the curvature? What is the ultimate fate of the Universe in this case?

If the density is less than the critical density, then $\Omega_0 < 1$, so the curvature is negative. Since there is not enough gravity to reverse the expansion, the Universe will continue expanding forever.

11.7. Explain how to use globular clusters to put a lower limit on the age of the Universe (see also Problem 8.19 in Chapter 8).

Globular clusters are made of stars that were all born at the same time. In addition, all the stars are approximately the same distance away, so that differences in their apparent magnitudes represent differences in their absolute magnitudes. Therefore, we can make an H-R diagram, plotting the magnitude of the stars versus their color or spectral type. From the location of the main sequence turn-off, we can find the age of the globular cluster. Since the globular clusters could not have been created before the Universe, the Universe must be at least as old as the oldest globular clusters.

11.8. Why were no significant amounts of elements heavier than lithium or beryllium formed in the Big Bang?

In the Big Bang, conditions were very hot. It was too hot for elements heavier than lithium to live very long (they were almost immediately torn apart by photons). Then the Universe cooled very quickly, over just a few hundred seconds. As soon as it was cool enough for the elements to survive, it was too cool to make them.

11.9. If a gas of temperature 3000 K emitted CMBR, why does the CMBR have the spectrum of a 2.728 K blackbody?

Since the radiation was emitted, space has expanded, red-shifting the radiation, so that it now represents a much cooler blackbody.

11.10. Astronomers observe several galaxies in the Local Group that actually approach the Milky Way. The Andromeda Galaxy is the largest of these. Does this mean that the Hubble Law is incorrect? Why or why not?

No, this does not disprove the Hubble Law. On small scales, such as the size of the Local Group, the behavior of space is dominated by gravity, rather than by expansion.

Life in the Universe

WHAT IS LIFE?

1. **Organization**. Life is organized and structured. This is true at all levels: molecular, cellular, organ, individual, etc. To maintain this organization requires the use of energy (metabolism).
2. **Reproduction**. Things that are alive reproduce.
3. **Responds**. Life responds to stimuli, such as excessive heat or cold, light or darkness, etc.

FAVORABLE CONDITIONS FOR LIFE

Life on Earth is based on amino acids, which are carbon-based molecules. Water seems also to be a crucial factor. Water is an excellent solvent. It can dissolve materials so that they can move and interact. Water is also a liquid at a wide range of temperatures (0–100°C), and it can act as a temperature regulator as it absorbs heat on evaporation and releases heat on condensation. Other liquids might serve these functions for life elsewhere, such as ammonia and methyl alcohol. Ammonia, however, has a smaller range of temperatures over which it is a liquid, and methyl alcohol has a low heat of vaporization, which means that methyl alcohol is not a good temperature regulator. Thus, abundance of carbon and water would indicate conditions favorable to life.

Planets and moons have lots of carbon. Mars may at one time have had liquid water, and so it seems a good place to look for fossils of life. Europa currently has lots of liquid water, so it might be a good place to look for current life.

MARS

The first attempt to find life on Mars, carried out by the Viking lander, was inconclusive. The lander carried out three tests, two of which had negative results, and the third of which had a positive result. This has been interpreted to mean that there is some sort of chemical (non-biological) process occurring that we do not understand.

On the other hand, in 1996, scientists announced that they may have found evidence of life in a Martian meteorite. Inside, they found organic compounds, which increased in number closer to the center of the meteorite. This is unusual. Organic compounds are often found in space, and in meteorites, but in most meteorites the number decreases near the center. Also, some tube-like structures were seen near the organic compounds that resemble fossils of living organisms. However, there are other, non-biological processes that can form similar tubules, and the organic compounds are the same as those found elsewhere.

CHAPTER 11 Cosmology

EUROPA

Voyager images of Europa show cracks and distortions that look much like the patterns seen in the cracked ice over the Arctic Ocean. There are objects that look like icebergs tilted on their sides, and Europa's density profile (mapped out by Voyager and Galileo) is consistent with a liquid water ocean under a crust of ice. The cracks may persist for centuries before the ice fills them in again, allowing sunlight to penetrate to the water below. Also, Europa is heated from within by tidal interactions, which also provide a source of energy.

ELSEWHERE

The Drake equation gives an estimate of the number of civilizations, N, in our galaxy that are able to communicate with others:

$$N = R_* f_p n_e f_l f_i f_c L$$

- R_* is the rate at which stars form in the galaxy: a few per year.
- f_p is the fraction of stars with planets. This number is probably about 0.1.
- n_e is the number of Earth-like planets (or moons) per planetary system where life can survive. This number is probably in the range from 0.1 to 1.
- f_l is the fraction of these planets on which life actually arises. Estimates range from 10^{-3} to 1.
- f_i is the fraction of life species on a planet that develop some form of intelligence. Estimates range from 10^{-6} to 1.
- f_c is the fraction of intelligent life forms that actually develop communication across space. On the Earth, this fraction is a number close to 1.
- L is the length of time the communicating society exists. For Earth, this number is at least 100 years. Guesses for the maximum of this number go up to 10^9 years.

Obviously, our limited knowledge of the factors in the Drake equation introduce substantial uncertainty in the result. The minimum number N must be 1, since we are here.

LOOKING FOR LIFE

From another star, the Sun would look like a variable radio source, since the Earth and the Sun would be indistinguishable from that distance. The signal would vary with a period of 24 hours, as the Earth rotates on its axis. So, we look for life in a similar way. The Arecibo Radio Telescope is used to look for periodic radio signals from stars.

The Arecibo Radio Telescope was also used to send signals to other stars, but the light travel time is so long that a response is not expected for at least several thousand years. It is important to note that the entire discussion above relates only to our own galaxy. Even if the probability of life in the Galaxy turns out to be small, the probability of life in the Universe may still be large, since there are so many galaxies.

Solved Problems

11.11. Suppose that R_* is 5/year, f_p is 0.5, n_e is 2, f_l is 1/1,000, f_i is 1/1,000, f_c is 1, and L is 1,000 years. What is N? Is this reasonable?

Drake's equation is the product of all these values,

$$N = R_* f_p n_e f_l f_i f_c L$$
$$N = 5 \cdot 0.5 \cdot 2 \cdot \frac{1}{1,000} \cdot \frac{1}{1,000} \cdot 1 \cdot 1,000$$
$$N = 5 \times 10^{-3}$$

Using these values, N is a completely unreasonable 5×10^{-3}. N must be at least 1 to account for the fact that humans exist.

11.12. Of the values in Problem 11.11, the most uncertain values are f_l and f_i. How much must these number be changed in order to make $N > 1$?

If each of these quantities increased to 1/10, then $N = 50$, which is greater than 1, and therefore acceptable in principle.

11.13. Proxima Centauri is the closest star to the Sun. It is 1.3 pc away. How long would it take for us to receive a reply to a message sent to Proxima Centauri?

The light would have to travel to Proxima Centauri, and return, so it would have to travel 2.6 pc all together. Using the relationship between time, distance, and velocity gives

$$t = \frac{d}{v}$$
$$t = \frac{2.6 \cdot 3 \times 10^{16} \, \text{m}}{3 \times 10^8 \, \text{m/s}}$$
$$t = 2.6 \times 10^8 \, \text{s}$$
$$t = 8.2 \, \text{yr}$$

It would take 8.2 years for a message to be sent and replied to from Proxima Centauri.

Supplementary Problems

11.14. Suppose that $H = 65 \, \text{km/s/Mpc}$, so that $\rho_c = 8 \times 10^{-27} \, \text{kg/m}^3$. You observe the density of the Universe to be $7.9 \times 10^{-27} \, \text{kg/m}^3$. What is the curvature?

Ans. Negative

CHAPTER 11 Cosmology

11.15. Suppose the density of the Universe is 10^{-26} kg/m^3. How many hydrogen atoms are in a box 1 m on a side?

Ans. Nearly 6

11.16. What is the peak wavelength of radiation produced by a 3,000 K blackbody? What type of radiation is this?

Ans. 9.66×10^{-7} m, infrared

11.17. What is the speed of protons in the early moments after the Big Bang, when the temperature was 10 billion K? (Use thermal speed from Chapter 1.)

Ans. 1.4×10^7 m/s

11.18. How much kinetic energy does the proton in Problem 11.17 have?

Ans. 1.7×10^{-13} J

11.19. When the proton in Problems 11.17 and 11.18 meets an antiproton, what is the frequency of the radiation production?

Ans. 4.6×10^{23} Hz

11.20. What is the wavelength of a typical photon produced by a gas of 1 billion K? What is the frequency of this photon?

Ans. 2.9×10^{-12} m, 1×10^{20} Hz

11.21. How much energy does the typical photon of Problem 11.20 have?

Ans. 6.85×10^{-14} J

APPENDIX 1

Physical and Astronomical Constants

Symbol	Value	Meaning or other name
π	3.1415926	pi
c	2.9979×10^8 m/s	speed of light in vacuum
G	6.67×10^{-11} m^3/kg/s^2	gravitational constant
h	6.6261×10^{-34} W·s^2 (W·s^2 = J·s)	Planck's constant
m_e	9.1094×10^{-31} kg	mass of electron
m_H	1.6735×10^{-27} kg	mass of hydrogen atom
σ	5.6705×10^{-8} W/m^2/K^4	Stefan-Boltzmann constant
k	1.3805×10^{-23} W·s/K	Boltzmann constant
M_{Earth}	5.9742×10^{24} kg	mass of Earth
M_{Sun}	1.9891×10^{30} kg	mass of Sun
R_{Earth}	6.378×10^6 m	radius of Earth (at equator)
R_{Sun}	6.9599×10^8 m	radius of Sun
L_{Sun}	3.8268×10^{26} W	luminosity of Sun
AU	1.496×10^{11} m	astronomical unit
pc	3.0857×10^{16} m	parsec
ly	9.4605×10^{15} m	light year

APPENDIX 2

Units and Unit Conversions

Symbol	Equivalent unit	Name, What does it measure?
nm	1×10^{-9} m	nanometer, length
μm	1×10^{-6} m	micrometer (micron), length
cm	1×10^{-2} m; 0.3937 inches	centimeter, length
m	3.28 feet	meter, length
km	1×10^{3} m; 0.6214 miles	kilometer, length
AU	1.496×10^{11} m	astronomical unit, length
ly	9.4605×10^{15} m	light year, length
pc	3.0857×10^{16} m; 3.26 ly; 206,265 AU	parsec, length
Mpc	10^{6} pc	megaparsec, length
kg	2.2046 pounds (on Earth)	kilogram, mass
yr	3.16×10^{7} s	year, time
M_{Earth}	5.9742×10^{24} kg	mass of Earth
M_{Sun}	1.9891×10^{30} kg	mass of Sun
R_{Earth}	6.378×10^{6} m	radius of Earth (at equator)
R_{Sun}	6.9599×10^{8} m	radius of Sun
L_{Sun}	3.8268×10^{26} W	luminosity of Sun

APPENDIX 3

Algebra Rules

SCIENTIFIC NOTATION

Scientific notation is a way of writing numbers in shorthand. For example, $300 = 3 \times 10^2$ (to make the number 300, multiply 3 by 10 twice). Combining numbers in scientific notation means following three rules.

1. **Adding and subtracting**. Numbers written in scientific notation can only be added and subtracted if the exponent on the 10 is the same. Then, simply add or subtract the numbers before the ×. For example,

$$3 \times 10^8 + 4 \times 10^8 = 7 \times 10^8$$
$$3 \times 10^8 - 4 \times 10^8 = -1 \times 10^8$$

and

$$3 \times 10^8 + 4 \times 10^9 = 3 \times 10^8 + 40 \times 10^8$$
$$= 43 \times 10^8$$
$$= 4.3 \times 10^9$$

2. **Multiplying**. Multiply the numbers before the × and add the exponents.

$$(3 \times 10^8) \cdot (4 \times 10^7) = 12 \times 10^{15}$$

3. **Dividing**. Divide the numbers before the × and subtract the exponents.

$$(3 \times 10^8)/(4 \times 10^7) = 0.75 \times 10^1$$
$$= 7.5$$

SIGNIFICANT DIGITS

The final answer should always have only as many significant digits as the measurement with the **least** number of significant digits.

$$(2.81 \times 10^2) \cdot (8 \times 10^5) = 2 \times 10^8$$

APPENDIX 3 Algebra Rules

ORDER OF OPERATIONS

Powers are performed first, then multiplications and divisions, and finally additions and subtractions. (Operations in parentheses are carried out first.)

$$5 + 6 \cdot 7 = 47$$
$$(5 + 6) \cdot 7 = 77$$
$$(5 + 6)^7 = 2 \times 10^7$$
$$5 + 6^7 = 3 \times 10^5$$

UNITS

Just as numbers which appear in both the numerator and denominator of a fraction cancel, so do units.

$$(N \cdot m/m) = N$$
$$(km/s) \cdot (Hz/km) = 1/s^2$$

LOGARITHMS

Any number can be written as 10^x, if we allow x to be a non-integer:

$$4 = 10^{0.6}$$

To invert this, use the log function on your calculator:

$$0.6 = \log(4)$$

or

$$42 = 10^{1.62}$$
$$1.62 = \log(42)$$

APPENDIX 4

History of Astronomy Timeline

Period	Dates	Who	What
Ancient	35,000 BC	Lascaux Caves	Include Sun/star symbols
	7,000 BC	Abris de las Vinas (Spain)	First known lunar phase diagram
	3,500 BC	Proto-Europeans	Began building megalithic stone structures such as Stonehenge
	3,000 BC	Babylonians/Egyptians	Identified constellations
	2,000 BC	Babylonians	Recorded motions of planets
		Babylonians/Egyptians	Identified ecliptic
	500 BC	Greeks	Widely understood that the Earth and Moon are spherical
	293–273 BC	Eratosthenes	Measured circumference of the Earth
	200 BC	Babylonians	Predicted lunar/solar eclipses
	200 BC	Babylonians/Egyptians	Clearly recognized precession of Earth's poles
Medieval	4th–11th century	Arabs and Persians	Intensive development of astronomy: star charts and catalogues, planets, and the Moon movement; better estimations of the Earth size and calendar improvement
	813	Al Mamon	Founded the Baghdad school of astronomy
	813	Ptolemy	*Mathematike Syntaxis* by Ptolemy is translated into Arabic as al-Majisti (Great Work), later called by Latin scholars *Almagest*
	903	Al-Sufi	Constructed his star catalog
	1054	Chinese astronomers	Observe supernova in Taurus (now this supernova remnant is known as the Crab Nebula (M1))
Renaissance	1543	Copernicus	Published *De Revolutionibus Orbium Coelestium* in which he provided mathematical evidence for the heliocentric theory of the Universe
	1572	Tycho Brahe	Discovered a supernova in the constellation Cassiopeia
	1576	Tycho Brahe	Founded the observatory at Uraniborg

APPENDIX 4 History of Astronomy Timeline

Period	Dates	Who	What
	1582	Pope Gregory XIII	Introduced the Gregorian calendar
	1595	David Fabricius	Discovered the long-period variable star in the constellation Cetus, named Mira Ceti
	1600 (Feb. 17)	Giordano Bruno	After 8 years of imprisonment, was charged with blasphemy, immoral conduct, and heresy for challenging the official church doctrine on the origin and structure of the universe and burned at the stake in Campo dei Fiori
	1603	Johann Bayer	Published his star catalog, *Uranometria*. Introduced the Bayer system of assigning Greek letters to stars—still widely used by astronomers
	1604	Kepler	Discovered supernova in Ophiuchus
	1608	Lippershey	Dutch spectacle maker invented the first telescope
	1609	Galileo	First used the telescope for astronomical purposes: discovered four Jovian moons, observed Lunar craters and the Milky Way
		Kepler	Announced first two laws of planetary motion
	1611	Galileo, Scheiner, and Fabricius	Observed sunspots
	1612	Peiresc	Discovered the Orion Nebula (M42)
	1619	Kepler	Published the Third Law of Planetary Motion in his *Harmonice Mundi* (*Harmony of the World*)
	1631	Kepler	Predicted the transit of Mercury, which was subsequently observed by Gassendi
	1632	Galileo	Published his Dialogue on the Two Chief World Systems—the discussion of Ptolemaic and Copernican hypotheses in relation to the physics of tides (the original version, *Dialogue on the Tides*, was licensed and altered by the Roman Catholic censors in Rome)
	1633	Galileo	Was forced by the Inquisition to recant his theories
	1639	Horrocks	Observed the transit of Venus
	1647	Hevelius	Published a map of the Moon
	1656	Huygens	Discovered the nature of Saturn's rings and Titan—the largest satellite of Saturn
	1659	Huygens	Observed markings on the planet Mars
	1666	Cassini	Observed the polar caps on Mars
	1668	Newton	Built the first reflecting telescope (Newtonian)
	1669	Montanari	Discovered the variable nature of Algol
	1671	Paris Observatory	Founded
	1675	Greenwich Observatory	Founded
		Romer	Measured the velocity of light
		Cassini	Discovered the main division in Saturn's rings
	1683	Cassini	Observed the zodiacal light
Modern	1687	Newton	Published *Philosophiae Naturalis Principia Mathematica* establishing the theory of universal gravitation
	1705	Halley	Predicted the return of Halley's comet in 1758
	1725	Flamsteed	The first Astronomer Royal of England, published his star catalog. He introduced star numbering in each constellation in order of increasing right ascension

APPENDIX 4 History of Astronomy Timeline

Period	Dates	Who	What
	1728	Halley	Discovered proper motion
		Bradley	Proposed the theory of the aberration of the fixed stars, including the aberration of light
	1729	Hall	Proposed the principle of the achromatic refractor
	1750	Wright	Speculated about the origin of the solar system
	1755	Kant	Proposed the hypothesis of the origin of celestial bodies
	1758	Palitzsch	Observed previously predicted Halley's comet return.
	1761	Lomonosov	Discovered the atmosphere of Venus
	1781	Messier	Searching for comets, Messier discovered dozens of deep sky objects (galaxies, nebulae, and star clusters) which he compiled in his catalog
	1781	Herschel	Discovered Uranus
	1784	Goodricke	Discovered the variable nature of Delta Cephei
	1789	Herschel	Built a telescope at Slough with a 48-in mirror. Using this telescope he resolved stars in different galaxies
	1796	Laplace	Proposed the Nebular Hypothesis of the origin of the solar system based on the theory of stellar evolution
	1801	Piazzi	Discovered the first asteroid, Ceres
	1802	Herschel	Announced the discovery of binary star systems
		Wollaston	Observed absorption lines in the solar spectrum
	1803		Fall of meteorites at L'Aigle. The explanation of the nature of meteorites is established
	1811	Olber	Proposed the theory of comet tails
	1814	Fraunhofer	Provided a detailed description of the solar spectrum
	1834	Bessel	Inferred that the irregularity of proper motion of Sirius is due to the presence of an invisible companion
	1837	Beer and Madler	Published the first accurate map of the moon
	1838	Bessel	Determined the distance of 61 Cygni. This was the first determination of a stellar distance
	1839–40	Draper	The first application of photography to astronomy—Draper photographed the Moon
	1842	Doppler	Discovered the Doppler effect
	1843	Schwabe	Described the sunspot cycle
	1846	Galle	Discovered the planet Neptune based on its position calculated by the French astronomer Leverrier
	1851	Foucault	Provided evidence for the rotation of the Earth by suspending a pendulum on a long wire from the dome of the Pantheon in Paris
	1859	Kirchoff	Interpreted the dark lines in the Sun's spectrum
	1859–62	Argelander	Published *Bonner Durchmusterung* (BD)—a catalog of over 300,000 stars
	1862	Clark	Discovered Sirius B based on calculations by Bessel
	1860–63	Huggins	Began the spectral analysis of stars
	1868	Jansen and Lockyer	Observed solar prominences
	1877	Hall	Discovered the Martian satellites Phobos and Deimos
		Schiaparelli	Observed the Martian canals

APPENDIX 4 . History of Astronomy Timeline

Period	Dates	Who	What
	1890	Lockyer	Announced his theory of stellar evolution
		Vogel	Discovered spectroscopic binaries
	1894	Percival Lowell	Founded the Flagstaff Observatory in Arizona
	1897	Yerkes Obs.	Founded
20th century	1900	Chaberlin and Moulton	Proposed the new theory of the solar system origin
	1905	Mt. Wilson Obs.	Established exclusively for the study of the Sun
	1905	Einstein	Proposed the basis of the Special Theory of Relativity, first described in his paper *On the Electrodynamics of Moving Bodies*
	1908	Hertzsprung	Described giant and dwarf stars
		Leavitt	Discovered the period–luminosity relation for Cepheids
	1911–14	Hertzsprung and Russell	Discovered the relationship between spectral type and absolute magnitude (H-R diagram)
	1914	Goddard	Began practical experiments with rockets
	1915	Adams	Discovered White Dwarfs (Sirius B)
	1916	Eddington	Proposed the first premises of the theory of stellar structure
		Einstein	Proposed his General Theory of Relativity
	1918	Shapley	Provided the first model of the Galaxy structure
	1918–24	Cannon	Published the fundamental catalog of star spectra
	1919	Barnard	Published the catalog of dark nebulae
	1920	Slipher	Discovered red shifts in the spectra of galaxies
	1923	Hubble	Proved that the galaxies lie beyond the Milky Way
	1926	Goddard	Fired the first liquid fuel rocket
	1927	Oort	Proved that the center of the galaxy lies in the direction of Sagittarius
	1929	Hubble	Discovered linear relationships between the galaxy distance and its radial velocity, the Hubble Law
	1930	Tombaugh	Discovered Pluto based on Lowell's predictions
	1931	Jansky	Discovered cosmic radio waves
	1937	Reber	Constructed the first radio telescope
	1937–40	Gamow	Proposed the first theory of stellar evolution
	1942	Strand	Speculated that 61 Cygni is attended by a planet.
	1944	Van de Hulst	Suggested that interstellar hydrogen must emit radio waves at 21.1 cm
	1946	Bay	Obtained the first radar images of the Moon
	1947	Ambarcumian	Discovered star associations
	1949	Hale 200-inch	Completed at Mount Palomar
	1951	Ewen and Purcell	Discovered the 21.1 cm hydrogen emission predicted by van de Hulst
	1951–54		Spiral structure of our galaxy determined
	1955		250-foot radio telescope at Jodrell Bank is completed
	1957	Russia	The first artificial satellite launched
	1958	USA	The first American satellite launched
	1959	Russia	Lunik I passes the Moon; Lunik II lands on the Moon

APPENDIX 4 History of Astronomy Timeline

Period	Dates	Who	What
	1961	Gagarin	The first man in space
	1962	Glenn	First American orbital flight
		Russia/USA	Planetary probes: Mars I (Russia) and Mariner II (USA)
			First galactic source of X-ray radiation (Sco X-1) detected
			First quasar (3C273) discovered
	1965	Penzias and Wilson	Discovered cosmic background radiation, providing direct evidence to support the Big Bang Theory
	1966	Russia/USA	First soft landing on the Moon (Luna 9—Russia and Surveyor I—USA). Russian probe lands on Venus
	1967	Bell, Hewish	Discovered pulsars
	1968	Apollo 8: Borman, Lovell, and Anders	First manned flight around the Moon
	1969	Apollo 11: Armstrong and Aldrin	July 20–21: First man on the Moon
	1970	Uhuru	Satellite Uhuru scans the sky in the X-ray range
	1971	Russia	First probes in orbit around Mars and first soft landing on Mars (Mars 3—Russia)
	1971	USA	First manned mechanical vehicle on the Moon (Apollo 15—USA)
	1972		Satellite Copernicus conducts spectroscopic ultraviolet observations of stars and interstellar matter with high resolution The first observations in gamma-radiation range Launch of Pioneer 10—the first probe to Jupiter (USA)
	1973	USA	First images of Jupiter transmitted from close vicinity by Pioneer 10
	1977		Discovery of Uranian rings
	1978		Discovery of Pluto's moon, Charon
	1980	USA	First images of Saturn and its rings transmitted from close vicinity by Voyager 1
	1983		InfraRed Astronomical Satellite scans the sky in the infrared
	1986		January 24: Voyager 2 approaches the planet Uranus January 28: Space shuttle Challenger disaster March: Space probes Vega 1, Vega 2, and Giotto pass near Halley's comet
	1987		February 23: Supernova 1987a in the Large Magellanic Cloud was visible to naked eye
	1988		Discovery of quasars 17 billion light years away
	1989		May 4: Magellan mission to radar map the surface of Venus August 24: Voyager 2 approaches the planet Neptune November 18: NASA launches Cosmic Background Explorer (COBE) satellite
	1990		April 24: space shuttle Discovery puts the Hubble Space Telescope into orbit December 5: the first picture (galaxy NGC 1232 in Eridanus) taken with Keck Telescope is published in the *Los Angeles Times*
	1991		April 5: Compton Gamma Ray Observatory (GRO) launched October: Galileo passes by the asteroid Gaspra

APPENDIX 4 History of Astronomy Timeline

Period	Dates	Who	What
	1992		April: the Hubble Space Telescope photographs in the Large Magellanic Cloud the hottest star ever recorded (temp. 360,000°F) April 24: COBE proves the existence of temperature fluctuations in the background radiation, which is strong evidence supporting the Big Bang theory. September 16: the discovery of the first object orbiting the Sun beyond the planet Pluto, in the Kuiper Belt September 25: NASA launches the Mars Observer spacecraft to study the atmosphere and surface of Mars October 31: the Vatican (Pope John Paul II) announce that the Catholic Church erred in condemning Galileo's beliefs
	1993		January 31: The Gamma Ray Observatory (GRO) detects the brightest burst of gamma rays ever recorded December: Astronauts aboard space shuttle Endeavour correct the defects in the Hubble Space Telescope
	1994		July 20: Comet Shoemaker-Levy crashes into Jupiter
	1995		December 7: Galileo reaches the planet Jupiter

INDEX

Absolute magnitude, 127
Absorption lines, 19
Accretion disks, 50
Active galactic nuclei, 196
Active galaxies, 196
Active region, 136
Albedo, 21
Altitude, 31
Amino acids, 214
Angles, measuring, 25
Angular diameter, 25
Angular momentum, 4
Angular resolution, 43
Annihilation of matter, 154
Antiparticles, 154
Apparent brightness, 127
Apparent magnitude, 127
Arcminute, 25
Arcsecond, 25
Asteroids, 99
Asymptotic Giant Branch (AGB), 160
Atmosphere:
  of Earth, 65
  of Jupiter, 78
  of Mars, 71
  of Mercury, 56
  of Moon, 67
  of Neptune, 80
  of Pluto, 103
  of Saturn, 79
  of Uranus, 80
  of Venus, 59
  planetary, 54
Aurorae, 63
Autumnal equinox, 31
Azimuth, 31

Balmer series, 129
Barred spirals, 190
Big Bang:
  background radiation from, 206
  time of occurrence, 204
Blackbody emission, 117
Black hole, 179
Blueshift, 20
Bombardment era, 51
Brightness:
  measuring, 127
  of stars, 127

Broad line regions, 198
Butterfly diagram, 140

Callisto, 83
Carbonaceous chondrites, 96
Celestial coordinate system, 31
Celestial equator, 31
Cepheid variables, 159
Chondrites, 96
Chondrules, 96
Chromosphere, 128
Circular velocity, 4
Clouds:
  of interstellar gas, 109
  on Jupiter, 78
  on Mars, 71
  on Neptune, 80
  on Venus, 60
Clusters of galaxies, 192
Collisions between galaxies, 192
Comet:
  defined, 91
  and meteor showers, 96
  origins of, 92
  parts of, 91
  properties of, 91
Condensation, 50
Continuous emission, 17
Convection zone, 137
Core:
  of Earth, 62
  of Jupiter, 78
  of Mars, 72
  of molecular clouds, 117
  of stars, 153
  of supernovae, 162
Corona, 142
Coronal holes, 142
Coronal mass ejections, 140
Cosmic background radiation, 19
Cosmological principle, 203
Cosmology, 203
Craters, 54
  on Mercury, 56
  on Moon, 67
Crescent phase, 34
Critical density, 208
Crust, 53
  of Earth, 62

Crust (*Cont.*)
  of Mars, 72
  of Moon, 69
Curvature of space, 208
Curved spacetime, 208

Dark matter, 186
Dark nebulae, 110
Day:
  lengthening of, 33
  lunar, 33
  sidereal, 32
  solar, 32
Declination, 31
Decoupling epoch, 208
Density:
  critical, 208
  and curvature, 208
  definition of, 2
Differential rotation, 140
Differentiation, 53
Diffusion, radiative, 137
Disks, 183
Distance:
  calculating, 25
  to galaxies, 191
  to stars, 126
Doppler effect:
  defined, 20
  and stellar spectra, 128
Doppler equation, 20
Dust:
  in comet tails, 91
  interstellar, 110
  in Seyfert galaxies, 198
Dwarfs, 156

Earth:
  atmosphere of, 65
  evolution of, 53
  interior of, 62
  properties of, 49
Eccentricity, 2
Eclipse:
  lunar, 35
  solar, 35
Ecliptic, 32
Eddington luminosity, 198
Electromagnetic radiation, 16

INDEX

Electromagnetic spectrum, 16
Electrons, 19
Ellipses, 2
Elliptical galaxies, 190
Elliptical orbits, 2
Emission lines, 19
Energy:
    gravitational potential, 5
    kinetic, 4
    and mass, 154
    in photons, 16
    produced in stars, 153
    and Stefan-Boltzmann law, 18
Energy level, 19
Equinox:
    autumnal and vernal, 31
    precession of, 35
Escape velocity, 4
Europa, 83
Event horizon, 179
Evolution of stars, 153
Expansion of universe, 204
Extinction by interstellar dust, 111

Filaments, solar, 140
Flares, solar, 140
Flat space, 208
Flatness problem, 209
Flux, 18
Foci, of ellipse, 2
Frequency, 15
Full phase, 34
Fusion:
    described, 153
    and energy, 154
    hydrogen, 153

Galactic bulge, 184
Galactic disk, 183
Galactic halo, 184
Galactic nucleus, 184
Galaxies:
    active, 196
    classification of, 189
    clusters of, 192
    distance to, 191
    elliptical, 190
    formation of, 192
    irregular, 191
    radial velocity of, 191
    rotation of, 185
    spiral, 189
Galilean satellites, 82
Gamma rays:
    creation of matter from, 154
    energy of, 16
    in stars, 154
Ganymede, 83

Gases:
    degenerate, 173
    density and pressure of, 13
    escape of, 13
    interstellar, 109
    neutral, 13
Giant molecular clouds, 107
Giants, 158
Gibbous phase, 34
Granulation of Sun, 138
Gravitational contraction, 117, 155
Gravity, 2
    effect on expansion, 205
    and space, 208
    and tides, 33
Greenhouse effect, 65

Helioseismology, 143
Helium:
    abundance, 164
    formation in primordial universe, 165
    fusion of, 153
    from hydrogen in stars, 153
Helium flash, 158
Hertzsprung-Russell diagram, 156
HII regions, 109
Horizon, 31
Horizon problem, 209
Hubble's constant, 204
Hubble's Law, 191, 203
Hubble time, 204
Hydrogen:
    absorption and emission, 129
    abundance, 164
    conversion to helium, 153
    21 cm line of, 110
Hydrostatic equilibrium, 126

Ideal gas law, 13
Impact cratering, 54
Inflation of universe, 209
Infrared radiation:
    defined, 16
    from interstellar dust, 110
Interstellar matter:
    dust, 110
    gas, 109
Interstellar reddening, 111
Inverse square law, 25
Io, 82

Jets, 198
Jovian planets, 76
Jupiter, 78

Kepler's Laws, 3
Kinetic energy, 4
Kuiper belt, 92

Latitude, 31
Law of gravitation, 2
Life, 214
Light:
    absorption and reflection of, 19
    measuring devices for, 44
    speed of, 16
Light curve, 101
Light-gathering power, 42
Line emission, 19
Linear diameter, 25
Lithosphere, 53
Local Group, 192
Long-period comets, 92
Luminosity:
    of active galactic nuclei, 196
    classifying, in stars, 130
    and mass, 131
    of stars, 127
Lunar eclipse, 35
Lunar phases, 33

Magnetic field:
    of Earth, 63
    of neutron stars, 176
    in star formation, 117
    of Sun, 139
Magnification of telescope, 44
Main-sequence stars:
    evolution of, 157
    lifetime of, 131
Mantle, 53
Maria, of Moon, 67
Mars, 71
Mass:
    defined, 1
    and energy, 154
    of stars, 131
    and stellar luminosity, 131
    of white dwarfs, 161
Mass–luminosity relation, 131
Massive stars:
    collapse of, 162
    life cycle of, 162
Matter, creation from radiation, 154
Mercury, 56
Meridian, 31
Metallic hydrogen, 78
Meteorites, 95
Meteoroids, 95
Meteors, 95
Milky Way, 183
Moon, 67

INDEX

Motion:
    proper, 128
    of stars, 128

Narrow line region, 198
Neap tides, 33
Nebulae:
    dark, 111
    HII regions, 109
    planetary, 161
    reflection, 111
Negative curvature, 208
Neptune, 80
Neutral gases, 13
Neutrinos, 137, 154
Neutron capture, 165
Neutron stars, 176
New phase, 34
Normal spirals, 189
North celestial pole, 31
Northern lights, 63
Nuclear reactions, 153, 155

Opacity, 156
Orbits, 3
Outgassing, 54

Parallax, stellar, 126
Penumbra, 139
Perihelion, 3
Period–luminosity relation, 159
Photons:
    creation from matter, 154
    effect on energy levels, 19
Photosphere, 128
Planetary nebulae, 161
Planetesimals, 50
Planets, 49, 77
Plasma, 13
Plasma tail of comet, 91
Plate tectonics, 53
Pluto, 103
Polarity:
    of Sun's magnetic field, 140
    of sunspots, 139
Positive curvature, 208
Positrons, 154
Precession, 35
Pressure, 13
Prominences, solar, 140
Proper motion, 128
Proton–proton chain, 153
Protostars, 117
Pulsars, 176

Quarter phase, 34
Quasars, 198

Radial velocity, 128
Radiant of meteor shower, 97
Radiation:
    absorption and reflection of, 19
    cosmic background, 206
    and matter, 154
    from quasars, 198
Radiation era, 206
Radio waves, 16
Recombination epoch, 208
Reddening, interstellar, 111
Red giant stars, 158
Red shift, 20
Reflecting telescopes, 42
Reflection nebulae, 111
Refracting telescopes, 42
Resolution of telescopes, 43
Right ascension, 31
Rings, 87
Rotation:
    differential, 140
    of Milky Way, 185
    of neutron stars, 176
    synchronous, 33
Rotation curve, 185
r-process, 165
RR Lyrae stars, 160

Saturn, 79
Scattering by interstellar dust, 111
Seasons, 34
Semimajor axis, 2
Short-period comets, 92
Small angle equation, 25
Solar eclipse, 35
Solar flares, 140
Solar prominences, 140
Solar system, origin of, 50, 77
Solar wind, 143
Solstices, 34–35
South celestial pole, 31
Spectra, 17
Spiral galaxies, 189
Spring tides, 33
s-process, 165
Stars:
    brightness of, 127
    distances to, 126
    energy generation in, 154
    formation of, 117
    main sequence, 131
    mass of, 131
    motions of, 128
    post-main sequence, 158
    red giant, 158
    spectra of, 128
    spectral class of, 128
    white dwarf, 161

Stefan-Boltzmann Law, 18
Stellar parallax, 126
Stony-iron meteorites, 96
Stony meteorites, 96
Structure problem, 210
S-type asteroids, 100
Summer solstice, 35
Sun:
    chromosphere of, 141
    corona of, 142
    interior of, 137
    photosphere of, 138
    sunspots in, 139
Sunspots, 139
Superclusters, 192
Supergranulation, 138
Supernovas, 162
Synchronous rotation, 33

Tails of comets, 91
Tectonics, 54
Telescopes, 42
Temperature:
    of gases, 13
    of stars, 128
    of terrestrial planets, 49
Terrestrial planets, 49
Tidal bulges, 33
Tidal forces, 33
Tides, 33
Titan, 85
Triton, 85
Type I supernovae, 163
Type II supernovae, 162

Ultraviolet radiation, 16
Universe:
    age of, 204
    contraction of, 209
    end of, 209
    expansion of, 204
    inflation of, 209
    measuring curvature of, 208
Uranus, 79

Van Allen belts, 63
Velocity:
    of atoms and molecules in a gas, 13
    circular, 4
    escape, 4
    of light, 16
    of orbiting bodies, 3, 4
Venus, 59
Vernal equinox, 31
Volcanism, 54
Volume, 1

INDEX

Waning, of Moon, 34
Wavelength, 15
Waves, electromagnetic, 15
Waxing, of Moon, 34
White dwarfs:
  age of, 174
  evolution of, 174

Wien's Law, 18
Winds:
  in AGB stars, 160
  on Jupiter, 78
  solar, 143
  in young stars, 118

X-rays, 16

Zenith, 31
Zero curvature, 208